What is Information?

Propagating Organization in the Biosphere, Symbolosphere, Technosphere and Econosphere

Robert K. Logan

Logan, Robert K.
What is Information? Propagating Organization in the Biosphere, Symbolosphere, Technosphere and Econosphere
Robert K. Logan.

Includes bibliographic references.
ISBN 978-1-60888-996-9

First edition published by DEMO Publishing, Toronto, Canada.
Printed on Demand (POD) in the United States of America by IngramSpark.
Access ebook editions or contact the editors at http://demopublishing.com.

Editors: Peter Jones and Greg Van Alstyne
Book design by Garry Ing & Greg Van Alstyne
Cover design by Greg Van Alstyne

DEMO stands for Design Emergence Media Organization.
DEMO Publishing presents new genres and formats of interdisciplinary design and media research. We seek to illuminate complex social and organizational systems, to foster foresight and innovation. All of these domains manifest emergence and demand continual updating of our systemic design thinking.

Contents

Bridging the gap by sapheron http://www.flickr.com/photos/sapheron/5265248965/ Attribution-ShareAlike

Preface

Our first monograph in the DEMO series is Professor Robert Logan's *What is Information?* Originally conceived for readership in information sciences and the McLuhan media ecology community, Bob's latest book was reviewed by a university publisher as a conventional scholarly monograph. Authorial titles are no longer sold to many scholars and libraries, and university publishers now decline most of these proposals. The editors (Peter Jones and Greg Van Alstyne) know Logan's body of work, and recognize his contributions as a unique, powerful interface of systemic thinking, media and language studies, and complexity. His work exemplifies the types of books we expect to publish in the Author imprint of reviewed monographs.

What is Information? is a unique title within information studies. It is strongly interdisciplinary, crossing information theory, systems theory, new media and cognitive linguistics. Therefore, it may carry provocative themes and insights that require of the reader a broader frame of reference than the known, narrow path. Among these interruptive inquiries is Bob's notion of there being different forms and frames of information in ecological contexts. Logan starts by denying Shannon's fundamental information theory a comprehensive reach, while respecting its boundaries. With help from Stuart Kauffman, he shows that biotic information—the instructions of life and reproduction—requires a different theory of information entirely from bit-oriented signal processing.

The book takes on the complex issue of defining information as a carrier of meaning versus signals processed by meaning-makers. Recovering the importance of MacKay's original contribution of the "distinction that makes a difference," Logan bridges information and media theory. If meaning is the coherence of organization, then information as meaning remains consistent with the notion of negative entropy. While media may shape the expression and meaning of meaning, it is information that signals the meaning of the medium. The power of language in developing symbols generates a constant source of meanings through information. To better distinguish these functions of "information" Dr. Logan relates information as a functional power of organization within four ecosystems: Biosphere, Symbolosphere, Technosphere, and Econosphere. The Biosphere gives rise to human cultures through information, and culture gives rise to the other three spheres. Information is the mediator of these spheres.

What is Information? benefits from a rich interactive collaboration with designers and the author. An editorially-directed design collaboration invested the title with a visually communicative expression that communicates and amplifies meaning beyond the text.

Dr. Logan is also Chief Scientist and one of the founders of Strategic Innovation Lab, and we disclose he is fully participating in the experimental function of this publication. The core publishing team—Garry Ing, Peter Jones, and Greg Van Alstyne—being members of an academic design lab, also aim to innovate publishing processes, by running studies on publishing models based on small volume, precisely targeted titles. We also hope to change design scholarship, if only in a small way, by innovating the meaning of scholarly design work. While this collaborative design was produced post-review, we plan to share new manuscripts (as appropriate) with our registered community for both review and possible contribution of design concept to enhance meaning.

About DEMO

We founded DEMO Publishing in 2012 as a research group within Strategic Innovation Lab (sLab) at OCAD University in Toronto. DEMO's mission is to launch new formats and genres of interdisciplinary design and media research currently unavailable in today's online and print media. DEMO, as *Design Emergence Media Organization*, presents theoretical work and design research underpinning innovation systems and behaviors. DEMO intersects design theories of media and the social ecosystem, conceptual media design, organizational and business design as communications (media), and foresight and innovation in complex systems. All of these domains manifest emergence and demand continual updating of our systemic design thinking.

Our Publications and Projects

DEMO products are organized around three imprints (or form factors), each a different approach to publication and a model for researching new publishing processes. These include:

- Scholarly author monographs
- Edited article collections
- Collaborative or sponsored research projects

We will advertise and sell online, socially and through the "colleges" that produce authors and reviewers.

Our Research Model

Strategic Innovation Lab (sLab) pursues research to explore and accelerate innovation futures, and as design researchers we recognized the dearth of design influence in scholarly publishing. While the top journals are excellent vehicles for the traditional article, they are not yet standing up for the interdisciplinary designer breaking new ground outside the known design disciplines. We also recognize that design scholarship does not offer publications as platforms for visual and expressive languages consistent with designerly values. Design journals and articles ought to reflect the cultures that define their forms of excellence.

We are not going to wait for publishers to change. DEMO can model the changes we foresee in the knowledge ecology. We will make DEMO books and collections available through online retail channels such as Amazon, our website, and group purchases.

Editorial and Authorship

To be clear to prospective authors, we do not expect DEMO to become a growing independent publishing business. It operates like a scholarly journal, a labour of care for thinkers and their ideas in the design communities. We are academics and working designers interested in learning from and sharing a more transparent and community-oriented approach to scholarly communication and field development.

We expect our first authors to be those active in our communities of practice and inquiry. Some authors write monographs, others research articles, and we find high value in graduate student research projects and working papers as well. Each of these types deserves a compatible form.

We encourage authors interested in proposing a title or submitting a manuscript to contact an editor directly and express the concept briefly in summary. We will exchange and consult among our board and respond with advice or response. We encourage notable and published scholars in these interdisciplinary fields to serves as editorial board members for DEMO. Please contact the Editors if interested via http://demopublishing.com.

Socialize this Book!

We believe readers and authors will communicate the discovered value. In this edition, we invite you to take up correspondence with author Bob Logan and the publishers through the channels detailed at the end of the book.

Foreword

Terrence Deacon

Information is probably the single most important factor shaping the beginning of twenty-first century social life. Without question, our current age is appropriately described as the "information age." Every aspect of human life is rapidly being invaded and restructured by information technology.

We are swimming in an ocean of recorded audio and video that can be literally part of one's apparel and one's constant companion. We rely on the ability to communicate at a moment's notice with our friends and acquaintances from almost anywhere in the world and at any time, using a cellular phone. We don't think twice about our individual power to send thoughts, images, personal opinions, or intimate diary notes to potentially millions of recipients in a few minutes, while sipping coffee in a wifi-equipped café. And we casually take advantage of the capacity to instantly access a large fraction of the written knowledge of the ages—from a digital library that is orders of magnitude larger than all but the largest libraries in the world. This provides instant access to the greatest literary works ever written, the latest findings of science or medicine, as well as answers to the trivia about movie stars, the latest hair styles, scores of recent football games, or the location of the nearest northern Italian restaurant in an unfamiliar city. None of this was ever imagined by even the most prescient futurists of just a generation ago. O brave new world, that has such creations in it!

But do we really understand what has happened to us in these few short decades? We now find ourselves scrambling to keep up with the flood of new information technologies that come to the market daily, but are we equally as attentive to the global and personal consequences of their cultural influence?

Does anyone have a clear perspective on how this is influencing our cultures, our identities, and our very thinking processes? Yes, there are innumerable new magazines and blogs surveying the rapidly shifting information technology landscape, but in this glut of info-talk is there anyone explaining what it is that is being processed, stored, mined, browsed, and corrupted?

Because information is not merely some tool that we can use or ignore, but is what also constitutes the very fabric of human identity and experience, these new information technologies almost effortlessly integrate into everyday life. Whereas a shovel or automobile remains physically separate from its users, the seemingly non-physical nature of information blends seamlessly into our everyday thoughts, perceptions, and beliefs. The very freedom from any fixed physical instantiation that makes information so fluid and sharable is also what provides it with remarkable stealth and influence. This easily blurred boundary between the churning of the information "out there" and what we imagine that we are and know and want "in here", gives these technologies the power to remake the very nature of our humanness.

As immersed as we are in this sea of information and its centrality to every facet of contemporary life one might naively assume that we (or someone) must have a pretty clear idea of exactly what information is. Wrong! We seem to know it when we see it, but when asked to define it or explain what it is, even CEOs of major IT companies and professors in philosophy or computer science programs seem to prevaricate. Or worse, they offer a standard technical definition that is hardly even a shadow of the familiar concept, and whose mathematical formalism promises far more insight into the workings of information than it delivers.

So, what *is* information? And why is it such an enormously difficult question to answer with any clarity and thoroughness? It is an ambitious book that sets out to answer this question, much less present an elaborate theory of how it has morphed into a seemingly independent universe of meanings, rituals, art-forms, values, and technologies since our ancestors first learned to talk. Who would attempt such a challenge?

A generation ago Marshall McLuhan helped a whole culture notice how the nature of the media we use to communicate with (from speech to print to radio to television) can have far more profound social consequences than does the content it conveys. His famous catch-phrase "The medium is the message" inverted what otherwise had seemed like common sense. Unlike his many predecessors McLuhan focused on the ground rather than the figure as he examined the cultural and epistemological influences of the introduction of writing, print, electronic media and other communicative innovations. Not surprisingly, Robert Logan comes by his interest in the deeper aspects of the problem of information in large part because of his past collaborations with McLuhan on topics like the nature of number or the origins of writing. As a result, his tendency to notice and explore the non-obvious properties of information can be seen as a natural extension of this approach to communication in general.

The first couple of chapters review the history of ideas about information. This historical prelude plays the critical role in explaining how the concept of information was transformed from a fundamentally mental concept to a technical term that has little to do with the original colloquial use. In the process of providing a precise formalization suitable for engineering purposes, the concept of information was denuded of any of its mentalistic functions. So whereas the new technical concept of information—that grew out of the work of Hartley, Shannon, Weaver and others—essentially made the information age possible, in return it has robbed this defining term of its core meaning.

Logan is interested in exploring this lost meaning. He recognizes the powerful contributions of this exactly quantifiable notion of information to the fields of computation and communication engineering. But he takes his mission to be to reintroduce into theory the very features that were excluded in this process. Beginning with a survey of those thinkers who argued against this reduction of information to mere logical media properties, such as Gordon Mackay, Gregory Bateson, and even Claude Shannon himself, Logan argues that it is necessary to recognize a number of distinct and more developed concepts of information.

So rather than arguing that a given concept of information is accurate or not, he instead begins by showing in which contexts one or another theory of information works or doesn't work. He then explores these different concepts of information and their wider implications. For Logan information is not one thing and what we call information depends on the context. In this way he attempts to rescue the current technical concept of information and its abandonment of the core defining features of information, by defining a number of higher-order concepts of information that reinstate the roles of meaning and function, and decouple it from specific medium properties.

To attempt to redefine information in a way that retains its functional as well as its logical features Logan turns to an approach suggested by systems thinker Stuart Kauffman. He relies heavily on a paper that he coauthored with Kauffman and others that argues that a fundamental feature of information is that it inevitably involves the propagation of organization. He refers to this paper (titled "Propagating Organization: An Enquiry") throughout the book with the abbreviation POE. This approach is based on two realizations: first that it takes work to produce constraints and constraints to do work, and second that information is always dependent on constraints. The technical engineering concept of information is based on a statistical understanding of the concept of constraint, and organization can be described in terms of constraint. Ever since James Clerk Maxwell introduced the scientific world to his little imaginary demon who used information about the velocities of individual gas molecules to reverse the Second Law of Thermodynamics people have assumed than in some way or other an increase of information is opposed to an increase in entropy, or disorder. Even though we know that in some sense this must be the case, showing exactly how has bedeviled researchers for the century and a half since then. This is an assumption that, while not completely explained, becomes a critical founding insight for the rest of the book. The remaining chapters offer probes of the implications of his novel way of conceiving information.

The core insight that runs through the book is a synthesis of two ways of conceiving of human thought and communication.

The first insight is the idea that mind—at least human mind—is not a phenomenon that is confined within a brain. The so-called "extended mind" hypothesis argues that what we consider a human mind is seamlessly integrated with the sea of communications it is embedded within. If, as Charles Sanders Peirce recognized over a century ago, cognition is a form of semiosis and each person is additionally engaged in semiotic exchanges with other minds, then in a real sense, no mind is an island. The boundary between your thought processes and mine is permeable, and the thoughts themselves exist in a distributed network that may extend across many individuals separated by thousands of miles and years.

The second insight is that this web of communication in many important respects exists and evolves and exhibits causal dynamics that persist in parallel with the physical-mechanical-biological world. This domain of meanings, purposes, and values has been called the "Symbolosphere" by John Schumann (2003).

These two insights are naturally compatible and are brought together in Logan's exploration of the wider implications of information. In addition, Logan suggests an interesting parallel with a Cartesian concept that in other respects he rejects. This is the infamous mind/body dualism that suggest that mental processes take place in a realm without physical substance or extension—*res cogitans*—and that the physical body is a materially constituted mechanism that is part of the physical world—*res extensa*. Cartesian dualism argues that the physical world is completely distinct from the realm of mind. Logan and Schumann reject the potential supernatural implications of this split and recognize that the realm of mind is very much part of the physical world. Yet they argue that the symbolic meanings and values generated by language nevertheless have a curious partial independence from any particular physical embodiment. They propose to resolve the Cartesian dilemma concerning how minds influence the physical world, in a practical way by simply identifying the Symbolosphere with Descartes' *res cogitans*. Although embodied in extended media, the meaningful contents of any communication are themselves not identical with this extended substrate. They are, as Deacon (2012) has noted, both physically absent from this immediate physical medium and yet an essential constituent, without which symbol tokens would be mere inert physical objects. So in this sense Logan and Schuman are prepared to treat the meanings and significance of these symbol tokens as separate from the *res extensa*. In this respect, this shadowy aspect of symbolic communication is consistent with a variant of Cartesian dualism, which Logan dubs *neo-dualism*. This provides a sort of compromise between a modern materialistic perspective and the classic mind/body dualism inherited from the Enlightenment.

Though I am not a fan of dualisms, I take Logan's notion of neo-dualism to be a sort of pseudo-dualism that trades on certain connotations implicit in the metaphysical analogy to classic dualism, while avoiding both the problem of explaining the windowless parallelism between the mental world and the physical world, and the seeming paradox of their interaction. This allows him to focus attention on the many features that distinguish communication from the merely mechanical features of the physical world. Indeed, he locates the epistemic cut between the material and meaningful at a juncture that Descartes himself would have felt comfortable with: with brain and body on one side and the world of symbols on the other.

But of course all forms of communication are at the same time meaning-making processes *and* physical processes. It's just that in many cases, such as in the symbolic communication provided by language, there is considerable flexibility with respect to this physical embodiment. Thus, the same linguistic meaning can be embodied in many different media, e.g., sound of voice, manual signs, hand-written scripts, print, or bits and bytes of computer memory. This doesn't make communication non-physical, just substrate-transferrable. There is always something that is extended in space and time that the communication requires in order to be realized: i.e. a medium. Nevertheless, the purpose of an action, the meaning of a word, the function of a tool, or the value of a work of art are at once dependent on a physical medium and something absent from that embodiment.

It was, of course, the genius of McLuhan to recognize the fundamental role played by the form of this medium. So it might at first appear that Logan has forgotten this essential insight. But despite his neo-Cartesian treatment of the various "spheres" of information processes that humans inhabit in addition to the physiosphere, biosphere, and their own biosemiotic and neurological processes, Logan remains quite solidly rooted in physical processes that for Descartes would have constituted the res extensa. His explorations of information processes in language, culture, technology, economics, and so forth, smoothly meld the meaningful with the physical, and are constantly focusing attention their physical causal influences.

It is likely that many readers will conclude that Logan doesn't exactly answer the question that is asked by the book's title. Providing a theory of information that formally explains the basis of reference and significance, and demonstrates the relationship between information and physical work remains a complex challenge for future research. To additionally link such an expanded theory of information with a theory of the evolution of language and an analysis of the evolutionary dynamics underlying cultural and technological change makes this an extremely ambitious project. Even if we are not provided with a full reformulation of the concept of information, Logan clearly demonstrates the many serious limitations in our current conceptions of information. The recognition that such a theory will need to be a component of a larger theory of the "propagation of organization', and not merely the reproduction and transmitting of bits of data, sets the stage for exploring the pragmatic aspects of symbolic communication. This provides him with a springboard for jumping into a far-reaching discussion of the many uniquely human modes of social-semiotic evolution that characterize our current historical era, from science and technology to computation and the internet.

Much of the remainder of the book highlights possibilities and challenges posed by these media and their different but related symbolic evolutionary processes, that only a more fully fleshed out theory of information can provide. Although neither the notion of the extended mind nor of the symbolosphere are mainstream science, this way of re-understanding media provides a fresh new perspective from which to view the unprecedented changes in the nature of human mentality wrought by being embedded in many levels of symbol-mediated information processing. We humans are the message being constantly reshaped in the image of the symbolic media we have and are creating. This is a book that clearly celebrates what it means to be a thoroughly symbolic species.

Terrence Deacon is a Professor of Anthropology in the Helen Wills Neuroscience Institute and a member of the Cognitive Science faculty at the University of California, Berkeley. He is the author of two important books, The Symbolic Species *and* Incomplete Nature.

Tom Jobim ao vivo em Montreal
YEAH JAZZ
SIX LANE ENDS
belén ille
Aurora
BACK TO SLOW FOOD
HOWLING WOLF
SPIRITUAL UNITY
PAT METHENY GROUP
Small Cars PostcardBook
1010 CHECK
ITALIAN
エヌ氏の遊園地

Attribution-NonCommercial-NoDerivs License

Chapter One

What is Information? Introductory Remarks to Frame the Investigation

What is information?—a fitting question given the importance of information and the central role it plays in the economic and cultural life at the beginning of the twenty-first Century.

It is said that we live in the Information Age, a claim that is hard to dispute given the ubiquity of the vast array of information technology (IT) at our disposal to generate, communicate, interpret and exploit information. We are surrounded by information thanks to computing and the digital "new media" such as the Internet, the Web, blogs, email, instant messaging, text messaging, cell phones, VOIP, Web cams, iPods, Blackberries, iPhones, virtual reality, virtual worlds, RFID or smart tags, nanotechnology and ubiquitous computing. In addition to the proliferation of these many informatic devices we also have to contend with the information explosion in the physical and biological sciences, engineering, social sciences, and humanities. In addition computing and IT have become the principal metaphor through which so much of our life and our world are understood as well as forming the underpinning of artificial intelligence (AI) and artificial life (AL). The ultimate information conceit, however, belongs to Edward Fredkin who insists that the universe is a computer and that life including human life is merely a program running on that computer (Hayles 1999, 240–42).

The irony of our total immersion in information and the central role it plays in our economic, social and cultural life is that for the most part we do not really have a clear understanding of exactly what information is. Information is not a simple straightforward concept but rather it is a very slippery notion used in many different ways and in many different contexts. Linguistically and grammatically the word information is a noun but in actuality it is a process and hence is like a verb. A consideration of the concept of information gives rise to a number of interesting questions, which we will explore in this study.

- Is there only one form of information or are there several kinds of information? In other words is information an invariant or a universal independent of its frame of reference or is it context dependent?
- What is the relationship of information to meaning, communication and organization?
- Is information a thing like a noun or a process like a verb?
- Is information material, a form of energy or is it just a pattern?
- Is information a uniquely human phenomenon or do non-human forms of life contain information also?
- What is the role of information in the propagation of life?
- What is the relationship of information to energy and entropy?
- What is the relationship of information to science?
- What is the relationship of information to media?

These are some of the questions we will address in this book as we try to flesh out our understanding of exactly what it is that we call information. We will consider the historic development of the concept of information to get a handle on the exact meaning of this thing or process that defines our age and is also the engine of economic growth. We trace the development of the concept of information from the earliest uses of the word to the beginning of information theory as formulated by Shannon and Wiener. We will also study the role of information in the four spheres of influence on human life, namely, the biosphere of living organisms, the symbolosphere, which consists of language, the human mind and all the products of the mind including culture; the technosphere of technology, and the econosphere of economics and government.

Investigations, Kauffman, 2000
Oxford University Press

The Background to "What is Information?": Three Previous Projects

This study grows out of three previous projects (described below) that have engaged my attention over the past few years. The actual question "what is information" arose out of a conversation with Stuart Kauffman and Robert Este in Ottawa immediately following the inaugural meeting of the Canadian Systems Biology Society. Stuart opened the conversation, as I best recall, by asking what is systems biology. I retorted by asking, isn't it about information in biotic systems. He responded by saying but what is information in a biotic system anyway? After a lively discussion of not more than an hour we concluded that Shannon information cannot properly describe biotic information. That conversation led to the first of three projects upon which this book is based and is entitled Propagating Organization: An Enquiry (Kauffman, Logan, Este, Hobill, Goebel and Shmulevich 2007) hereafter referred to as POE. In this enquiry we showed that biotic or instructional information is quite different than Shannon information and is related to the constraints that allow a living organism to convert free energy into work that allows it to operate its metabolism and replicate itself thus propagating its organization. The results of POE are summarized in Chapter 2.

Stuart Kauffman
Photo by Teemu Rajala

It should be mentioned that the concept of propagating organization was originally formulated by Kauffman (2000) in his book *Investigations* and that it will play a key role in this investigation. In fact it was by generalizing Kauffman's notion of propagating organization to language, culture, science, technology, and economics that I began to realize that there were many different forms of information in addition to Shannon and biotic information.

Although the question "What is Information?" arose out of that original conversation with Kauffman and Este I should also mention that the idea for the simple title of this book and this study, *What is Information?*, was very much influenced by Irwin Schrödinger's (1944) famous and highly influential book *What is Life?*

A second project that has contributed to my thoughts about the nature of information has been my ongoing work in media ecology and linguistics dating back to my collaboration with Marshall McLuhan (McLuhan and Logan 1977) and includes a study of the impact of the alphabet on the development of Western civilization in *The Alphabet Effect* (Logan 2004a), an attempt to understand the origin of human communications and language in *The Extended Mind* (Logan 2007) and its evolution ranging from speech, writing and mathematics to science, computing and the Internet in *The Sixth Language* (Logan 2004b). These results will be reviewed in **Chapters Three and Four** where we also examine the nature of the human mind. The motivation for understanding the human mind is that the original meaning of information pertained to giving a form to the mind. It is also the case that it is by the agency of the mind that information is formulated, communicated and received. We will therefore examine the nature of the human mind and its relationship to language and culture as developed in the Extended Mind model for the emergence of language and culture (Logan 2007). One of the key results from this work that is pertinent to addressing the question, "what is information?" is the way in which it was shown that verbal language, culture, technology and economics can be treated as though they were living organisms because of the way in which they evolve, have agency and represent emergent phenomena.

Erwin Schrödinger

The third project of relevance to understanding the nature of information arises from John Schumann's (2003a & b) formulation of the notion of the symbolosphere reviewed in **Chapter 5** and the collaboration Schumann and I (Logan and Schumann 2007) developed by combining his notion of the symbolosphere and my ideas from the Extended Mind model (Logan 2006a, 2007) to develop a neo-dualistic representation of reality. This work is reviewed in **Chapter 6**. The neo-duality representation suggests that reality consists of two basic elements: i. physical elements with extension occupying the physiosphere or what Descartes called res extensa and ii. symbolic elements like language, culture and mind, which are without extension and occupy what Schumann defined as the symbolosphere. The symbolosphere corresponds to Descartes' res cogitans but is agnostic with respect to the notions of God, soul and spirit. Logan (2006b) has extended the neo-duality concept to include media, science, music and the fine arts, which is also reviewed.

The Sixth Language, Logan, 2000

Stoddart Pub

The reviews of these past projects that I have sprinkled throughout this book are provided as the background for this study of the nature of information and have been collected in this volume for the convenience of those readers not familiar with some of my past work. The new material in this book is based on a number of explorations I have made and essays that I have written over the past five years all of which have related in one way or another to the notion of information, a concept or notion, which I find most puzzling. The fundamental questions that have intrigued me for the past five years are of the form "What IS X" where X is information, language, communications, culture, the book, mind, altruism, science, and life. Hopefully I have shed some light on these questions and the nature of information as it is used in such diverse fields of study as linguistics, communications, computer science, knowledge management, physics, biology, and cybernetics. I cannot claim expertise in any of these fields with the possible exception of physics where I earned my Ph.D. As an "intellectual tourists" and a genuine interdisciplinarian, however, I hope to have shed some light on all of these fields.

The Organization of This Book

Having described the three previous projects, which gave rise to this study we now turn to a description of the remaining chapters of this book.

Chapter Two introduces some of the issues associated with understanding the nature of information. The chapter begins with an etymological analysis of the term information and a history of the use of the word based on entries in the Oxford English Dictionary. Next we trace the history of the concept of information including Shannon's (1948) formulation of information theory, Weiner's (1948 and 1950) formulation of cybernetics, and the criticisms and limitations of Shannon information. We also explore the relationship of information to thermodynamics and entropy and argue, as have many physicists before us, that information and entropy are opposites and not parallel as suggested by Shannon.

Next we then turn to the use of information in the biological sciences and extensively review the article POE (Kauffman et al. 2007) where we have shown that Shannon information fails to describe biotic information. We also have shown that information is not an invariant but depends on the frame of reference or context in which it is used. We illustrate this latter point by examining the relationship of information to materiality and meaning in both biotic and symbolic information systems. We then show that there exists a link between information and organization in biotic systems and in the various aspects of human culture including language, technology, science, economics and governance. We end Chapter 2 by discussing whether a living organism like a human being is information or flesh and what is the relationship of information and flesh.

In **Chapter Three** we examine the origin and evolution of human language and its relationship to communication and information. The chapter reviews three previous studies, namely:

1. *The Alphabet Effect* (Logan 2004a) which posits that the phonetic alphabet, codified law, monotheism abstract science, and deductive logic first arose in the narrow geographic zone between the Tigris Euphrates river system and the Aegean Sea between 2000 BCE and 500 BCE among cultures that were trading and interacting with each other. This hypothesis was developed to help explain why abstract science began in the West despite the fact that most technology originated in ancient China.
2. *The Sixth Language* (Logan 2004b) which posits that language is both a medium of communication and an informatic tool and that speech, writing, mathematics, science, computing and the Internet form an evolutionary chain of languages.
3. *The Extended Mind* (Logan 2007, Chapters. 1–12 which posits that language emerged as the bifurcation from percept based mental processes to concept based thinking as a way of dealing with the complexity of hominid life due to tool making, the control of fire, the need to live in large social settings to take advantage of the hearth, large scale hunting and gathering and non-verbal mimetic communication needed to coordinate these activities. We also review the hypothesis developed by Christiansen (1994) and Deacon (1997) in which they posit that language may be treated as a living organism, an obligate symbiont that evolved so that it could be easily learned by young children. Their hypothesis obviates the need to invoke Chomsky's theory that a Lan-

guage Acquisition Device and the Universal Grammar was hard wired into the human brain to explain why young children learn language automatically.

In **Chapter Four** we examine the relationship of culture, organization and information making use of (Logan 2007, Chapters 2–4). We show that Christiansen's argument that language may be regarded as an organism can be extended to culture. We posit that culture as an organism evolved in such a way as to be easily learned and that as a result given the universality of human cognitive structures we should not be surprised by the universality of human culture as has been documented by Donald Brown (1991).

In **Chapter Five** we combine the results of Propagating Organization: An Enquiry (Kauffman et al. 2007) with the notion developed in Chapters Three and Four that language, culture, technology, economics and governance and science can be treated as organisms that evolve, propagate their organization and represent emergent phenomena. We also show that all of these human information systems also behave like living organisms with respect to three properties that Kauffman (2000) identified in *Investigations*, namely like living organisms

i. they constantly probe the Adjacent Possible,
ii. they maximize their variety and hence obey Kauffman's putative fourth law of thermodynamics, and
iii. they are self-constructing systems

In **Chapter Six** we examine the intersection of emergence theory and the concept of duality within the context of information and propagating organization that is materially instantiated in the case of the biosphere and that is not materially instantiated in the case of the symbolosphere. We show that the conflict that Clayton (2004) suggests exists between emergence and duality is easily resolved by introducing the notion of neo-duality (Logan and Schumann 2005) described above. We also point out the differences between Cartesian duality and Logan-Schumann neo-duality.

In **Chapter Seven** we describe the information content of the four spheres that directly influence the human condition, namely, the biosphere, the symbolosphere, the technosphere and the econosphere. We then compare the way in which the components of the four spheres, namely, living organisms, language and culture, technologies, and economic and governmental organizations:

1. contain information,
2. emerge and evolve,

3. develop their agency
4. are open to energy and information, and
5. enter into symbiotic relationships both within their own sphere and with those that reside in the other spheres.

In **Chapter Eight** we examine the relationship of information, knowledge, science and logic with a focus on two topics. The first is the role of information in knowledge management. The second topic treated in the section, "What is Science?," describes the limitations of science. We present a linguistic analysis and a formal mathematical proof, the Non-probativity Theorem, based on Popper's criteria of falsifiability for a scientific proposition to show that science cannot prove the truth of any proposition but can only formulate hypotheses that continually require empirical verification for every new domain of observation. A number of historical examples of how science has had to modify theories and/or approaches that were thought to be absolutely unshakable are presented including the shift in which linear dynamics is now considered the anomaly and non-linear dynamics the norm. Complexity and predictability are shown to have a complementarity like that of position and momentum in the Heisenberg uncertainty principle. The relationship of complexity and predictability is also similar to that of completeness and logical consistency within the context of Gödel's Theorem.

In **Chapter Nine** we examine the future of the book in the context of digital information. We conclude that despite some predictions of the obsolescence of the book its future looks bright. We describe the potential future of the evolution of the book describing the SmartBook system in which the convergence of the codex book and the e-book using a RFID smart tag results in a reading system that is readable, searchable, networkable and smart.

In **Chapter Ten** we examine the origin and nature of the non-verbal forms of information and communication inherent in the artistic expression through music, dance and the plastic arts of painting, sculpture, and photography. We also consider the connection between verbal language and artistic expression, which we believe is due to secondary perception, i.e. the perception influenced by verbal language and conceptual symbolic thought.

In this book, we address a number of interesting questions of a philosophical nature, which are worthy of our consideration despite the fact that any answers to them will be highly speculative. We show that despite all our efforts to understand the nature of information it is still a mysterious and somewhat ambiguous notion.

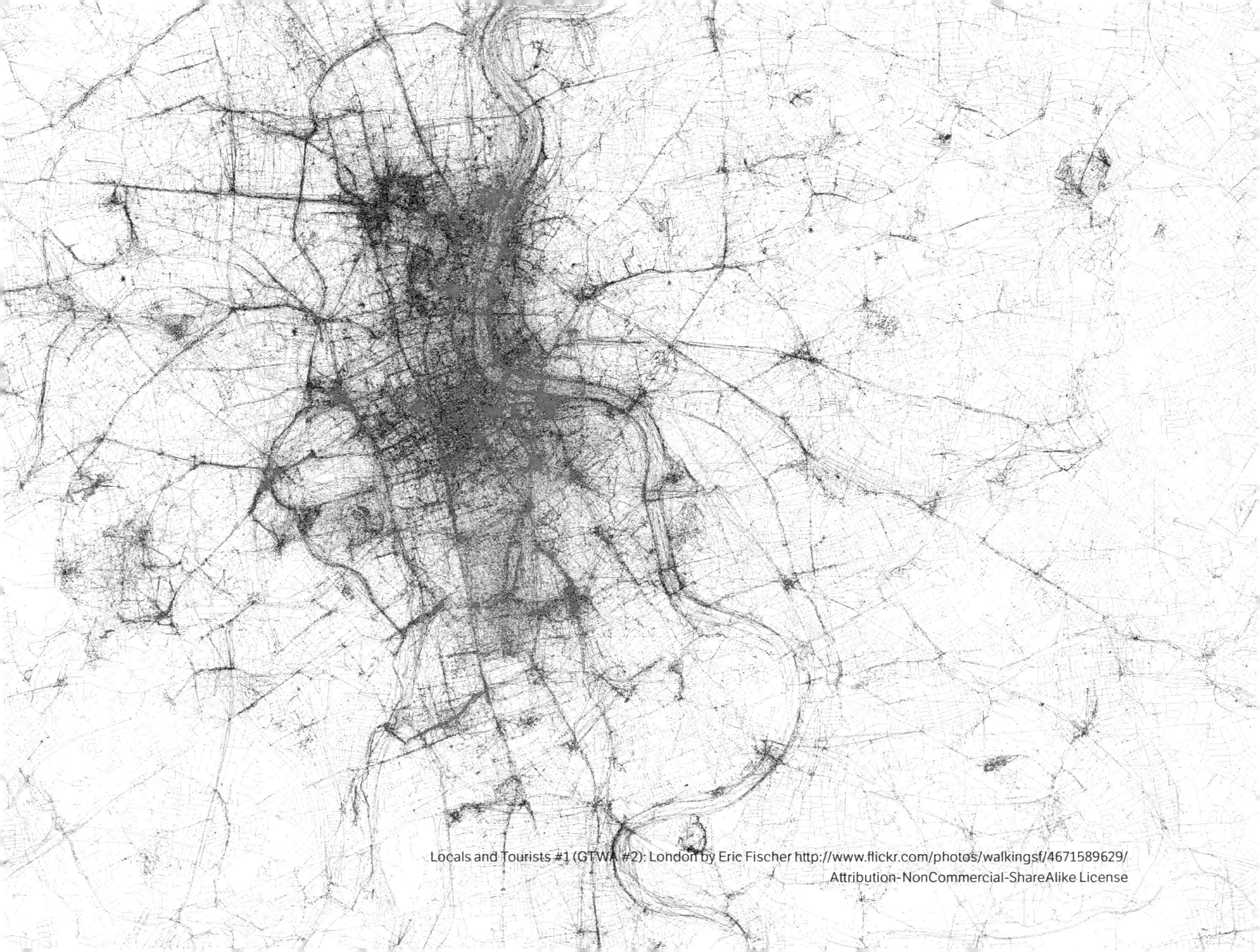

Locals and Tourists #1 (GTWA #2): London by Eric Fischer http://www.flickr.com/photos/walkingsf/4671589629/

Chapter Two

What is Information? Why is It Relativistic? & What is Its Relationship to Materiality, Meaning & Organization?

Juxtaposing these five modern definitions of information we begin to see the issues that we face in developing an understanding of information.

> *We have represented a discrete information source as a Markoff process. Can we define a quantity, which will measure, in some sense, how much information is 'produced' by such a process, or better, at what rate information is produced?*
> *—Shannon (1948)*

We see that Shannon's definition of information is a purely mathematical notion totally devoid of meaning or context. Bernd Frohmann (2004, 103) in his book *Deflating Information* referring to Shannon and Weaver's work writes, "Their interpretation of information is theoretically and mathematically rigorous, but it does not construe information in representational terms. Famously, and to some its analysis of information in terms of signal-to-noise ratios avoids the idea of meaning altogether." With Wiener's cybernetic concept of information we see that information takes on a functional role. The MacKay and Bateson definitions address their principal critique of

Shannon information, namely that it does not deal with meaning. Finally the last definition of Kauffman et al. (2007) deals with the materiality of biotic information and its relationship to organization, two features missing from Shannon information. These five quotes embrace the issues of information's materiality, meaning and relationship to organization that we will address in this chapter. We will show that the definition of information developed by Shannon that is commonly used in information theory only begins to scratch the surface of this complex phenomenon.

Claude Shannon

We begin by considering the historic development of the concept of information from its earliest usages in English to its formal formulation by the reputed father of information theory, Claude Shannon (1948), and the use of information in cybernetics by Norbert Weiner (1948, 1950). We then study the critiques of Shannon's formulation of information by Donald M. MacKay (1969) and Gregory Bateson (1973) who insist that information is more than a number of bits but that it also entails meaning. We also examine the relationship of information, energy and entropy arguing, as have many physicists before us, that information and entropy are opposites and not parallel as suggested by Shannon.

We then examine the way that information, which had always been associated with the human mind was introduced into biology by those considering the way genetic information is transmitted from one generation to another and by those considering the transmission of signals in living organisms.

We then review the work of Kauffman et al. (2007) that demonstrated that Shannon information cannot describe the information contained of a living organism. This work led to the introduction of the notion of the relativity of information and the realization that what we consider to be information depends on the context of where and how it is being generated and used.

Next we will examine the relationship of information to meaning and materiality within information theory, cybernetics and systems biology. And finally we examine the link between information and organization showing that in biotic systems that information and organization are intimately linked. We will also identify a similar link between information and organization in the various aspects of human culture including language, technology, science, economics and governance.

We conclude the chapter by discussing to what extent living organisms including humans are just flesh and to what extent they are information.

Origins of the Concept of Information

We begin our historic survey of the development of the concept of information with its etymology. The English word information according to the Oxford English Dictionary (OED) first appears in the written record in 1386 by

Chaucer: "Whanne Melibee hadde herd the grete skiles and resons of Dame Prudence, and hire wise informacions and techynges." The word is derived from Latin through French by combining the word inform meaning giving a form to the mind with the ending "ation" denoting a noun of action. This earliest definition refers to an item of training or molding of the mind. The next notion of information, namely the communication of knowledge appears shortly thereafter in 1450. "Lydg. & Burgh *Secrees* 1695 Ferthere to geve the Enformacioun, Of mustard whyte the seed is profitable."

The notion of information as a something capable of storage in or the transfer or communication to something inanimate and the notion of information as a mathematically defined quantity do not arise until the 20th century.

The OED cites two sources, which abstracted the concept of information as something that could be conveyed or stored to an inanimate object:

> **1937** *Discovery* Nov. 329/1 The whole difficulty resides in the amount of definition in the [television] picture, or, as the engineers put it, the amount of information to be transmitted in a given time.
> **1944** *Jrnl. Sci. Instrum.* XXI. 133/2 Information is conveyed to the machine by means of punched cards.

The OED cites the 1925 article of R.A. Fisher as the first instance of the mathematization of information:

> What we have spoken of as the intrinsic accuracy of an error curve may equally be conceived as the amount of information in a single observation belonging to such a distribution.... If p is the probability of an observation falling into any one class, the amount of information in the sample is $S\{(\partial m/\partial\theta)^2/m\}$ where $m = np$, is the expectation in any one class [and θ is the parameter] (Fisher 1925).

Another OED entry citing the early work of mathematizing information is that of R.V.L. Hartley (1928, 540) "What we have done then is to take as our practical measure of information the logarithm of the number of possible symbol sequences." It is interesting to note that the work of both Fisher and Hartley foreshadow Shannon's concept of information, which is nothing more than the probability of a particular string of symbols independent of their meaning.

Shannon and the Birth of Information Theory

Despite the early work of Fisher and Hartley cited above the beginning of the modern theoretical study of information is attributed to Claude Shannon (1948), who is recognized as the father of information theory. He defined information as a message sent by a sender to a receiver. Shannon worked at Bell Labs and wanted to solve the problem of how to best encode information that a sender wished to transmit to a receiver. Shannon gave information a numerical or mathematical value based on probability defined in terms of the concept of information entropy more commonly known as Shannon entropy. Information is defined as the measure of the decrease of uncertainty for a receiver. The amount of Shannon information is inversely proportional to the probability of the occurrence of that information, where the information is coded in some symbolic form as a string of 0s and 1s or in terms of some alphanumeric code. Shannon (1948, 392–94) defined his measures as follows:

> We have represented a discrete information source as a Markoff process. Can we define a quantity, which will measure, in some sense, how much information is 'produced' by such a process, or better, at what rate information is produced? Suppose we have a set of possible events whose probabilities of occurrence are $p_1, p_2,..., p_n$. These probabilities are known but that is all we know concerning which event will occur. Can we find a measure of how much 'choice' is involved in the selection of the event or of how uncertain we are of the outcome? If there is such a measure, say $H(p_1, p_2,..., p_n)$ ... we shall call $H = - p_i \log p_i$ the entropy of the set of probabilities $p_1..., p_n$... The quantity H has a number of interesting properties, which further substantiate it as a reasonable measure of choice or information.

A story is told that Shannon did not know what to call his measure and von Neumann advised him to call it entropy because nobody knows what it means and that it would therefore give Shannon an advantage in any debate (Campbell 1982, 32). This choice was criticized by Wicken (1987, 183) who argued that in science a term should have only one meaning. Schneider and Sagan (2005) referring to the use of the term entropy in both thermodynamics and information theory also suggests Shannon's use of the term is confusing when they wrote: "There is no simple correspondence between the two theories."

The Relationship of Information and Entropy

Understanding the efficiency of a steam engine through thermodynamics led Clausius to the idea of entropy as a measure of the mechanical unavailability of energy or the amount of heat energy that cannot be transformed into usable work. He referred to it in German as Verwandlungsinhalt, which may be translated roughly into English as "transformation content". Clausius then coined the term entropy deriving the root tropy from the Greek word trope (τροπή) meaning transformation. He added the prefix en because of the close

Leó Szilárd

association he felt that existed between energy and entropy. One can therefore roughly translate entropy from its etymology as energy transformation. Clausius felt the need to define entropy because the energy of the universe is conserved but its entropy is constantly increasing.

The relationship between entropy and probability is due to the work of Boltzman from his consideration of statistical mechanics, which is an alternative way of looking at thermodynamics. He showed that the entropy of a gas is proportional to the logarithm of W where W is the number of microstates of the gas that yield identical values of the thermodynamic variables of pressure, temperature and volume. The formula he derived, namely, that $S = k \ln W$ where k is the Boltzman constant is what inspired Shannon to call his expression for the measure of the information content of a message "information entropy" despite the difference in sign and the fact that the proportionality constant or Boltzman constant has the physical dimensions of energy divided by temperature.

The relationship between entropy and information as developed by physicists arose from a consideration of Maxwell's demon and is quite opposite to the one proposed by Shannon. Maxwell in 1867 postulated a gedanken experiment in which a demon standing in a doorway between two rooms filled with gas would allow only fast moving molecules to pass from one room to another so as to create a temperature difference in the two rooms from which usable work could be extracted in violation of the second law of thermodynamics. Leo Szilard in 1929 analyzing the problem that Maxwell's Demon presented showed that to obtain the information he needed the demon caused an increase of entropy elsewhere such that the net entropy did not decrease. He suggested that the demon is only able to temporarily reduce entropy because it possesses information, which is purchased at the cost of an increase in entropy. There is no violation of the Second Law because acquisition of that information causes an increase of entropy greater than the decrease of entropy represented by the information. As a result of Szilard's analysis one must conclude that entropy and information are opposite. He also pointed out that the net energy gained by the demon was not positive because of the energy cost in obtaining the information by which the demon selected the fast moving molecules and rejecting the slow moving ones. Since the information was purchased at the cost of an increase in entropy the information has an effective net negative entropy. Following Szilard, Gilbert N. Lewis (1930, 573) also saw an inverse relationship between information and entropy. He wrote, "Gain in entropy always means loss of information, and nothing more." Schrödinger (1944, 71–72) first explicitly introduced the notion of negative entropy:

> Every process, event, happening—call it what you will; in a word, everything that is going on in Nature means an increase of the entropy

Norbert Wiener

> of the part of the world where it is going on. Thus a living organism continually increases its entropy—or, as you may say, produces positive entropy—and thus tends to approach the dangerous state of maximum entropy, which is death. It can only keep aloof from it, i.e. alive, by continually drawing from its environment negative entropy—which is something very positive as we shall immediately see. What an organism feeds upon is negative entropy. Or, to put it less paradoxically, the essential thing in metabolism is that the organism succeeds in freeing itself from all the entropy it cannot help producing while alive (Chapter 6).

Both Wiener (1950) and Brillouin (1951) both adopted Shannon's definition of information and its relation to entropy with the one exception of its sign, likely influenced by the arguments of Szilard (1929) and Schrödinger (1944). Wiener wrote,

> Information is "negative entropy"; it expresses purpose (1948).

> Messages are themselves a form of pattern and organization. Indeed, it is possible to treat sets of messages as having entropy like sets of states in the external world. Just as entropy is a measure of disorganization, the information carried by a set of messages is a measure of organization. In fact, it is possible to interpret the information carried by a message as essentially the negative of its entropy, and the negative logarithm of its probability. That is, the more probable the message, the less information it gives (39).... This amount of information is a quantity which differs from entropy merely by its algebraic sign and a possible numerical factor (1950, 129).

Brillouin (1951) also argued that a living system exports entropy in order to maintain its own entropy at a low level. Brillouin used the term negentropy to describe information rather than negative entropy.

The reason that Wiener and Brillouin consider entropy and information as opposites or regard information as negative entropy follows from the tendency in nature for systems to move into states of greater disorder, i.e. states of increased entropy and hence states for, which we have less information. Consider a system, which is in a state for which there is a certain finite number of possible configurations or microstates all of which are equivalent to the same macro state. The tendency of nature according to the second law of thermodynamics is for the number of microstates that are equivalent to the macrostate of the system to increase. Because there are more possible microstates as time increases and we do not know which particular microstate the system is in, we know less about the system as the number of possible microstates increases. It therefore follows that as the entropy increases the amount of information we have about the system decreases and hence entropy is negative information or vice-versa information is the negative of entropy. In other words the second law of thermodynamics tell us that when system A evolves into system B that system B will have more possible redundant or equivalent micro states than system A and hence we know less about system B than system A since the uncertainty as to which state the system is in has increased.

Wiener and Brillouin relate information to entropy with a negative sign whereas Shannon uses a positive sign. Hayles (1999, 102) notes that although this difference is arbitrary it had a significant impact. Observing that Shannon used the positive sign she also noted that "identifying entropy with information can be seen as a crucial crossing point, for this allowed entropy to be reconceptualized as the thermodynamic motor driving systems to self-organization rather than as the heat engines driving the world to universal heat death." For Wiener, on the other hand she wrote, "life is an island of negentropy amid a sea of disorder (ibid.")."

Despite the difference in the sign of information entropy assigned by Shannon and Wiener, Shannon was heavily influenced by Wiener's work as indicated by the way Shannon (1948) credits Wiener for his contribution to his thinking in his acknowledgement: "Credit should also be given to Professor N. Wiener, whose elegant solution of the problems of filtering and prediction of stationary ensembles has considerably influenced the writer's thinking in this field." Shannon also acknowledge his debt to Wiener in footnote 4 of Part III:

> Communication theory is heavily indebted to Wiener for much of its basic philosophy and theory. His classic NDRC report, *The Interpolation, Extrapolation and Smoothing of Stationary Time Series* (Wiley, 1949), contains the first clear-cut formulation of communication theory as a statistical problem, the study of operations on time series. This work, although chiefly concerned with the linear prediction and filtering problem, is an important collateral reference in connection with the present paper. We may also refer here to Wiener's *Cybernetics* (Wiley, 1948), dealing with the general problems of communication and control.

MacKay's Counter Revolution: Where is the Meaning in Shannon Information?

According to Claude Shannon (1948, 379) his definition of information is not connected to its meaning. Weaver concurred in his introduction to Shannon's A Mathematical Theory of Communication when he wrote: "Information has 'nothing to do with meaning' although it does describe a 'pattern'." Shannon also suggested that information in the form of a message often contains meaning but that meaning is not a necessary condition for defining information. So it is possible to have information without meaning, whatever that means.

Not all of the members of the information science community were happy with Shannon's definition of information. Three years after Shannon proposed his definition of information Donald Mackay (1951) at the 8th Macy Conference argued for another approach to understanding the nature of information. The highly influential Macy Conferences on cybernetics, systems theory, information and communications were held from 1946 to 1953 during which Norbert Wiener's newly minted cybernetic theory and Shannon's information theory were discussed and debated with a fascinating interdisciplinary team of famous scholars which also included Warren McCulloch, Walter Pitts, Gregory Bateson, Margaret Mead, Heinz von Foerster, Kurt Lewin and John von Neumann. MacKay argued that he did not see "too close a connection between the notion of information as we use it in communications engineering and what [we] are doing here ... the problem here is not so much finding the best encoding of symbols ... but, rather, the determination of the semantic question of what to send and to whom to send it." He suggested that information should be defined as "the change in a receiver's mind-set, and thus with meaning" and not just the sender's signal (Hayles 1999b, 74). The notion of information independent of its meaning or context is like looking at a figure isolated from its ground. As the ground changes so too does the meaning of the figure.

Shannon whose position eventually prevailed defined information as the pattern or the signal and not the meaning. The problem with MacKay's definition was that meaning could not be measured or quantified and as a result the Shannon definition won out and changed the development of information science. The advantage that Shannon enjoyed over MacKay by defining information as the signal rather than meaning was his ability to mathematize information and prove general theorems that held independent of the medium that carried the information. The theorizing that Shannon conducted through his combination of electrical engineering and mathematics came to be known as information theory. It is ironic that the OED cites the first use of the term "information theory" as that of MacKay's who used the term in a heading in an article he published in the March 1950 issue of the Philosophical Magazine.

Shannon's motivation for his definition of information was to create a tool to analyze how to increase the ratio of signal to noise within telecommunications. People that shared MacKay's position complained that Shannon's definition of information did not fully describe communication. Shannon did not disagree—he "frequently cautioned that the theory was meant to apply only to certain technical situations, not to communication in general (ibid., 74)." He acknowledged that his definition of information was quite independent of meaning, however, he conceded that the information that was transmitted over the telecommunication lines he studied often had meaning as the following quote from his original paper written at the Bell Labs indicates:

> The fundamental problem of communication is that of reproducing at one point either exactly or approximately a message selected at another point. Frequently the messages have meaning; that is they refer to or are correlated according to some system with certain physical or conceptual entities. These semantic aspects of communication are irrelevant to the engineering problem. The significant aspect is that the actual message is one **selected** from a set of possible messages. The system must be designed to operate for each possible **selection**, not just the one that will actually be chosen since this is unknown at the time of design. If the number of messages in the set is **finite** then this number or any monotonic function of this number can be regarded as a measure of the information produced when one message is chosen from the set, all choices being equally likely.
> (Shannon 1948 – bolding is mine)

I ask the reader to note that Shannon requires the number of possible messages to be finite as this will be a critical concern when we examine biotic information. I admire Shannon's frankness about his definition of information, which he devised to handle the engineering problems he faced. He was quite clear that his definition was not the unique definition of information but merely one definition of information suited for his engineering requirements. In the abstract to his paper, The Lattice Theory of Information Shannon (1953) wrote,

> The word "information" has been given many different meanings by various writers in the general field of information theory. It is likely that at least a number of these will prove sufficiently useful in certain applications to deserve further study and permanent recognition. It is hardly to be expected that a single concept of information would satisfactorily account for the numerous possible applications of this general field. The present note outlines a new approach to information theory, which is aimed specifically at the analysis of certain communication

> problems in which there exist a number of information sources simultaneously in operation.

What I find extraordinary is that his definition of information limited in scope by his own admission became the standard by which almost all forms of information were gauged. There have been some slight variations of Shannon information like Kolmogorov information used to measure the shortest string of 0s and 1s to achieve a programming result or represent a text on a computer or a Turing machine. But despite these small variations Shannon information has been accepted as the canonical definition of information by all except for a small band of critics.

I have purposely bolded the term selected and selection in the above quote of Shannon to highlight the fact that Shannon's definition of information had to do with selection from a pre-determined set of data that did not necessarily have any meaning. MacKay used this selective element of Shannon information to distinguish it from his own definition of information, which, unlike Shannon, incorporates meaning explicitly. He also had to defend his definition from the attack that it was subjective.

> Mackay's first move was to rescue information that affected the receiver's mindset from the 'subjective' label. He proposed that both Shannon and Bavelas were concerned with what he called 'selective information,' that is information calculated by considering the selection of message elements from a set. But selective information alone is not enough; also required is another kind of information that he called 'structural.' Structural information indicates how selective information is to be understood; it is a message about how to interpret a message—that is, it is a metacommunication (Hayles 1999a, 54–55).

Structural information must involve semantics and meaning if it is to succeed in its role of interpreting selective or Shannon information. Structural information is concerned with the effect and impact of the information on the mind of the receiver and hence is reflexive. Structural information has a relationship to pragmatics as well as semantics where pragmatics tries to bridge the explanatory gap between the literal meaning of a sentence and the meaning that the speaker or writer intended. Shannon information has no particular relation to either semantics or pragmatics. It is only concerned with the text of a message and not the intentions of the sender or the possible interpretations of the receiver.

Part of the resistance to MacKay information was that its definition involved subjectivity, which orthodox scientists could not abide in their theories. Rather than deal with the fact that the exchange of information among humans involves a certain amount of subjectivity proponents of Shannon information theory chose to ignore this essential element of information and communications. Taken to its logical conclusion this attitude would limit science to study those areas that do not involve subjectivity, which would forever condemn linguistics and the other social sciences to non-scientific analysis. Rule out subjectivity in science or social studies and social science becomes a contradiction in terms.

This raises the question of whether subjectivity can be studied scientifically. I would suggest that an approach that parallels quantum physics is needed. Just as the measurement of sub-atomic particles changes their behavior and requires a quantum mechanic representation that includes the Heisenberg Uncertainty principle, something similar is required for a science of the subjective—something I would call quantum rhetoric. What is the study of communications and media ecology after all but the study of how one set of subjective humans communicates with another set of subjective humans. Shannon successfully exorcised the subjectivity from communications, which was fine for his engineering objectives. I totally respect Shannon because he always warned that his definition was not intended to be a theory of communications. My problem is with those that misuse his work and over extend it.

Information: The Difference that Makes a Difference

Although Shannon's notion of information divorced from meaning became the central theme of information theory MacKay's counter-revolution was not without some effect and resulted in a slight shift in the way information was regarded. No doubt the reader is familiar with Gregory Bateson (1973, 428) famous definition of information as **"the difference that makes a difference."** Buried in this one-liner is the notion that it is the meaning of the information that makes the difference. Although Bateson gets credit for this idea it is likely that he was influenced by Donald MacKay who is thought to have said **"information is a distinction that makes a difference"** This quote is attributed by many authors to MacKay's (1969) book *Information, Mechanism and Meaning* published four years before the appearance of Bateson's one-liner but no written form of this saying by MacKay has been found. Bateson, MacKay and Shannon were all participants in the Macy conferences so Bateson was quite familiar with MacKay's ideas. The use of the term "distinction" in MacKay's one-liner is more closely tied to the idea of "meaning" than the term "difference". It is ironic that MacKay who pointed out the shortcomings of Shannon information, was the first to use the term "information theory" and was the first to point out that the importance of information is its meaning and the fact that it makes a difference. MacKay is certainly a scholar who made a difference and he deserves more credit and attribution than he usually receives.

Another one line definition of information that incorporates the notion of its meaning is this one by Ed Fredkin which I would put in a league with Mackay and Bateson's one-liners. "The meaning of information is given by the processes that interpret it." This is a very interesting definition because it explicitly incorporates the notion that information depends on context.

If information is the distinction (McKay) or the difference (Bateson) that makes a difference then if there is no distinction or no difference then there can be no information. This would mean chaos or random numbers contain no information because there is no difference or distinction in one part of the stream of numbers as opposed to another part of the stream because of a lack of organization. This is opposite to the conclusion of Shannon who claims that a stream of random numbers contains the maximum information possible. While it is true each element is different from the next and is a complete surprise it is also true that the overall pattern of chaos and randomness is the same and hence there is no distinction nor is there any difference in the stream of random numbers. A gas, which remains uniformly at the same temperature, pressure and volume, is constantly changing but one cannot make a distinction between the gas at one moment and the gas at another moment. There is no difference in the way the gas behaves at these different moments. The only information one can discern about the gas is its volume, pressure and temperature, which is unchanging. No work can be done by this gas. If, however, in this volume of gas there is a temperature differential then work can be extracted from the gas and there is information in the gas by virtue of the way in which the temperature differential is organized. This raises the question of whether or not organization is information, a point we will return to later in this chapter once we have dealt with the nature of information in biotic systems.

Information in Biotic Systems

We have seen that as early as 1925 the notion of information as an abstraction was first introduced by Fisher (1925) and formalized by Shannon (1948). It was not long after this development that biologists also began to talk about information. The OED cites the first uses of the term in biology in 1953:

> **1953** J. C. ECCLES *Neurophysiol. Basis Mind* i. 1 We may say that all 'information' is conveyed in the nervous system in the form of coded arrangements of nerve impulses.
> **1953** WATSON & CRICK in *Nature* 30 May 965/2 In a long molecule many different permutations are possible, and it therefore seems likely that the precise sequence of the bases is the code which carries the genetical information.

R.A. Fisher

The use of information in this context was not the mathematization of information as was done by Fisher and Shannon but rather information was thought of qualitatively as something capable of being transferred or communicated to or through a living organism or stored in a living organism in the form of a sequence of nucleic acids.

Life as Propagating Organization

Stuart Kauffman (2000) defined an autonomous agent (or living organism) acting on its own behalf and propagating its organization as a collective autocatalytic system carrying out at least one thermodynamic work cycle. The relationship of the information found in living organisms to the kind of information treated in Shannon information theory was not clear even though a lot of attention has been given in recent times to the notion of information in biotic systems by those pursuing systems biology and bioinformatics. It was to examine this relationship that a group of us undertook a study to understand the nature and flow of information in biotic systems. This led to an article entitled Propagating Organization: An Enquiry (POE) authored by Kauffman, Logan, Este, Goebel, Hobill and Shmulevich (2007) in which we demonstrated that Shannon information could not be used to describe information contained in a biotic system. We also showed that information is not an invariant independent of its frame of reference.

In POE we argued that Shannon's (1948) classical definition of information as the measure of the decrease of uncertainty was not valid for a biotic system that propagates its organization. The core argument of POE was that Shannon information "does not apply to the evolution of the biosphere" because Darwinian preadaptations cannot be predicted and as a consequence "the ensemble of possibilities and their entropy cannot be calculated (Kauffman et al. 2007)." Therefore a definition of information as reducing uncertainty does not make sense since no matter how much one learns from the information in a biotic system the uncertainty remains infinite because the number of possibilities of what can evolve is infinitely non-denumerable. I remind the reader that in making his definition that Shannon specified that the number of possible messages needed to be finite.

Instead of Shannon information we defined a new form of information, which we called instructional or biotic information,

> not with Shannon, but with constraints or boundary conditions. The amount of information will be related to the diversity of constraints and the diversity of processes that they can partially cause to occur. By taking this step, we embed the concept of information in the ongoing processes of the biosphere, for they are causally relevant to that which

> happens in the unfolding of the biosphere. We therefore conclude that constraints are information and ... information is constraints....
> We use the term **"instructional information"** because of the instructional function this information performs and we sometimes call it **"biotic information"** because this is the domain it acts in, as opposed to human telecommunication or computer information systems where Shannon information operates (ibid.).

A living organism is an open system, which von Bertalanffy (1968, 149) "defined as a system in exchange of matter with its environment, presenting import and export, building-up and breaking-down of its material components." Instructional or biotic information may therefore be defined as the organization of that exchange of energy and matter.

In POE we argued that constraints acting as instructional information are essential to the operation of a cell and the propagation of its organization.

> The working of a cell is, in part, a complex web of constraints, or boundary conditions, which partially direct or cause the events which happen. Importantly, the propagating organization in the cell is the structural union of constraints as instructional information, the constrained release of energy as work, the use of work in the construction of copies of information, the use of work in the construction of other structures, and the construction of further constraints as instructional information. This instructional information further constrains the further release of energy *in diverse specific ways,* all of which propagates organization of process that completes a closure of tasks whereby the cell reproduces (ibid.).

In POE we associated biotic or instructional information with the organization that a biotic agent is able to propagate. This contradicts Shannon's definition of information and the notion that a random set or soup of organic chemicals has more Shannon information than a structured and organized set of organic chemicals found in a living organism.

> The biotic agent has more meaning than the soup, however.
> The living organism with more structure and more organization has less Shannon information. This is counterintuitive to a biologist's understanding of a living organism. We therefore conclude that the use of Shannon information to describe a biotic system would not be valid. Shannon information for a biotic system is simply a category error. A living organism has meaning because it is an autonomous agent acting on its own behalf.

> A random soup of organic chemicals has no meaning
> and no organization (ibid.).

The key point that was uncovered in the POE analysis was the fact that Shannon information could be defined independent of meaning whereas biotic or instructional information was intimately connected to the meaning of the organism's information, namely the propagation of its organization. Thus we see organization within a system as a form of information, which is a much more dynamic notion of information than Shannon information which is merely a string of symbols or bits.

According to Shannon's definition of information a set of random numbers transmitted over a telephone line would have more information than the set of even numbers transmitted over the same line. Once 2, 4, 6, 8, 10, 12 was received the receiver, who is assumed to be a clever person, would be able to correctly guess that the rest of the numbers to follow the sequence would be the set of even numbers. The random numbers have no organization but the even numbers are organized so the mystery of the relevance of Shannon information deepens as one must counter-intuitively conclude that information and organization can be at cross-purposes in Shannon's scheme of things.

This argument completely contradicts the notion of information of a system biologist who would argue that a biological organism contains information. It is by virtue of this propagating organization that an organism is able to grow and replicate, as pointed out by Kauffman (2000) in *Investigations*. From the contradiction between Shannon and biotic information we already have a hint that there is possibly more than one type of information and that information is not an invariant like the speed of light in relativity theory, which is independent of its frame of reference. We also see that perhaps Shannon's definition of information might have limitations and might not represent an universal notion of information. After all Shannon formulated his concept of information as information entropy to solve a specific problem namely increasing the efficiency or the signal to noise ratio in the transmission of signals over telecommunication lines.

The Relativity of Information

Robert M. Losee (1997) in an article entitled A Discipline Independent Definition of Information published in the Journal of the American Society for Information Science defines information as follows:

> Information may be defined as the characteristics of the output of a process, these being informative about the process and the input. This discipline independent definition may be applied to all domains, from physics to epistemology.

The term information, as the above definition seems to suggest, is generally regarded as some uniform quantity or quality, which is the same for all the domains and phenomena it describes. In other words information is an invariant like the speed of light, the same in all frames of reference. The origin of the term information or the actual meaning of the concept is all taken for granted. If ever pressed on the issue most contemporary IT experts or philosophers will revert back to Shannon's definition of information. Some might also come up with Bateson definition that information is the difference that makes a difference. Most would not be aware that the Shannon and Bateson definitions of information are at odds with each other. Shannon information does not make a difference because it has nothing to do with meaning; it is merely a string of symbols or bits. On the other hand, Bateson information, which as we discovered should more accurately be called MacKay information, is all about meaning. And thus we arrive at our second surprise, namely the relativity of information. Information is not an invariant like the speed of light, but depends on the frame of reference or context in which it is used.

We discovered in our review of POE that Shannon information and biotic or instructional information are quite different. Information is not an absolute but depends on the context in which it is being used. So Shannon information is a perfectly useful tool for telecommunication channel engineering. Kolmogorov (Shiryayev 1993) information, defined as the minimum computational resources needed to describe a program or a text and is related to Shannon information, is useful for the study of information compression with respect to Turing machines. Biotic or instructional information, on the other hand, is not equivalent to Shannon or Kolmogorov information and as has been shown in POE is the only way to describe the interaction and evolution of biological systems and the propagation of their organization.

Information is a tool and as such it comes in different forms just as screwdrivers are not all the same. They come in different forms, slot, square, and Philips—depending in what screw environment they are to operate. The same may be said of information. MacKay identified two main categories of information: selective information not necessarily linked to meaning and structural information specifically linked to meaning. Shannon information was formulated to deal with the signal to noise ratio in telecommunications and Kolmogorov information was intended to measure information content as the complexity of an algorithm on a Turing Machine. Shannon and Kolmogorov information are what MacKay termed selective information. Biotic or instructional information, on the other hand, is a form of structural information. The information of DNA is not fixed like Shannon selective information but depends on context like MacKay structural information so that identical genotypes can give rise to different phenotypes depending on the environment or context.

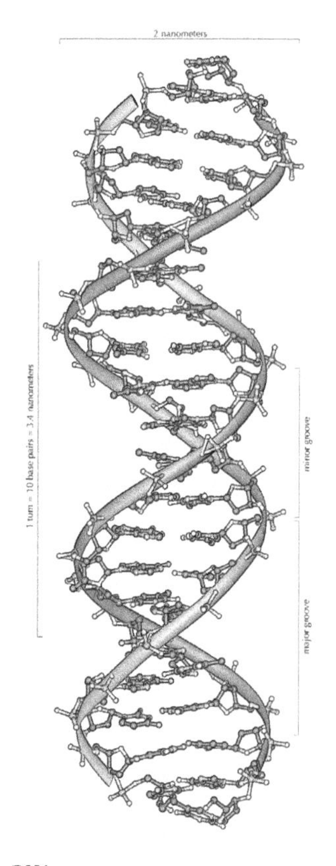

DNA

As MacKay and Bateson have argued there is a qualitative dimension to information not captured by the Shannon Weaver quantitative model nor by Kolmogorov's definition. Information is multidimensional. There is a quantitative dimension as captured by Shannon and Kolmogorov and a qualitative one of meaning as captured by MacKay and Bateson but one can think of other dimensions as well. In responding to a communication by Joseph Brenner on the Foundations of Information (FIS) listserv I described the information that he communicated as stimulating, provocative and enjoyable. Brenner cited the following Kolmogorov definition of information as "any *operator*, which changes the distribution of probabilities in a given set of events." Brenner's information changed the distribution of my mental events to one of stimulation, provocation and enjoyment and so there is something authentic that this definition of Kolmogorov captures that his earlier cited definition of information as "the minimum computational resources needed to describe a program or a text" does not. We therefore conclude that not only is there a relativistic component to information but it is also multidimensional and not uni-dimensional as is the case with Shannon information.

Although we introduced the notion of the relativity of information in POE we were unaware at the time of the formulation of a similar idea long ago by Nicholas Tzannes (1968). He "wanted to define information so that its meaning varied with context ... [and] pointed out that whereas Shannon and Wiener define information in terms of what it is, MacKay defines it in terms of what it does (Hayles 1999a, 56)." Both Shannon and Wiener's form of information is a noun or a thing and MacKay's form of information is a verb or process. We associate instructional or biotic information with MacKay as it is a process and not with Shannon because DNA, RNA and proteins are not informational "things" as such but rather they catalyze "processes" and actions that give rise to the propagation of organization and hence the transmission of information—information with meaning at that. Put simply instructional information is structural information as the root of the word in<u>struct</u>ional reveals.

Another distinction between Shannon information and biotic or instructional information as defined in POE is that with Shannon there is no explanation as to where information comes from and how it came into being. Information in Shannon's theory arrives deus ex machina, whereas biotic information as described in POE arises from the constraints that allow a living organism to harness free energy and turn it into work so that it can carry out its metabolism and replicate its organization. Kauffman (2000) has described how this organization emerges through autocatalysis as an emergent phenomenon with properties that cannot be derived from, predicted from or reduced to the properties of the biomolecules of which the living organism is composed and hence provides an explanation of where biotic information comes from.

Information and Its Relationship to Materiality and Meaning

N. Katherine Hayles

O, that this too too solid flesh would melt
—Shakespeare's Hamlet (Act 1, Scene 2)

Where is the wisdom we have lost in knowledge?
Where is the knowledge we have lost in information?
—TS Eliot

Where is the meaning we have lost in information?
—RK Logan

To drive home the point that information is not an invariant but rather a quantity that is relative to the environment in which it operates we will now examine the relationship of information to materiality and meaning drawing on the work and insights of Katherine Hayles (1999a & b). She points out that although information is used to describe material things and furthermore is instantiated in material things information is not itself material. "Shannon's theory defines information as a probability function with no dimension, no materiality, and no necessary connection with meaning. It is a pattern not a presence (Hayles 1999a, 18)."

The lack of a necessary connection to meaning of Shannon information is what distinguishes it from biotic information. Biotic information obviously has meaning, which is the propagation of the organism's organization. Information is an abstraction we use to describe the behavior of material things and often is sometimes thought of as something that controls, in the cybernetic sense, material things.

Hayles (1999a) traces the origin of information theory to cyberneticians like Wiener, von Forester and von Bertalanffy and telecommunication engineers like Shannon and Weaver. She points out that they regarded information as having a more primal existence than matter. Referring to the information theory they developed she wrote: "It (information theory) constructs information as the site of mastery and control over the material world."

She further claims, and I concur, that Shannon and cybernetic information is treated as separate from the material base in which it is instantiated. Wiener (1961, 132), for example, wrote in his book *Cybernetics, or Control and Communication in the Animal and the Machine* that "information is information, not matter or energy" (1961, 132). The question that arises is whether or not there is something intrinsic about information or is it merely a description of or a metaphor for the complex patterns of behavior of material things. Does information really control matter or is information purely a mental construct based on the notion of human communication through symbolic language, which in turn is a product of conceptual thought as described in Logan (2006 & 2007) and the next chapter?

While it is true that the notion of information as used by the cyberneticians like Wiener, von Forester and von Bertalanffy and that used by Shannon and Weaver influenced each other and in the minds of many were the same they are actually quite different from each other. The notion of information as the master or controller of the material world is the view of the cyberneticians beginning with Wiener (1950): "To live effectively is to live with adequate information. Thus, communication and control belong to the essence of man's inner life, even as they belong to his life in society."

For communication engineers information is just a string of symbols that must be accurately transmitted from one location, the sender, to another location, the receiver. Their only concern is the accuracy of the transmission with the relationship to the meaning of the information being meaningless to their concerns. If we consider the relationship of information and meaning for the moment then there is a sense in which the cybernetician's notion of information has meaning as a controller of the material realm whereas Shannon information has no relationship as such to meaning. In fact one can question if Shannon's used the correct term "information" when he described $H = - p_i \log p_i$ as the measure of "information". The quantity H he defined is clearly a useful measure for engineering in that it is related to the probability of the transmission of a signal—a signal that might or might not contain meaning. It is my contention that a signal without meaning is not truly information. I agree with MacKay and Bateson that to qualify as information the signal must make a difference, as is also the case with the way Wiener defines information in the context of cybernetics. Sveiby reports that Shannon himself had some second thoughts about the accuracy of his use of the term 'information':

> Shannon is said to have been unhappy with the word "information" in his theory. He was advised to use the word "entropy" instead, but entropy was a concept too difficult to communicate so he remained with the word. Since his theory concerns only transmission of signals, Langefors (1968) suggested that a better term for Shannon´s information theory would therefore perhaps be "signal transmission theory" (from the following Web site visited on 9/9/07: http://sveiby.com/portals/0/articles/Information.html#Cybernetics).

I find myself in agreement with Langefors that what Shannon is analyzing in his so-called information theory is the transmission of signals or data. It is consistent with some of my earlier work in the field of knowledge management and collaboration theory, in part inspired by the work of Karl Erik Sveiby, where Louis Stokes and I developed the following definitions of data, information, knowledge and wisdom:

- Data are the pure and simple facts without any particular structure or organization, the basic atoms of information,
- Information is structured data, which adds meaning to the data and gives it context and significance,
- Knowledge is the ability to use information strategically to achieve one's objectives, and
- Wisdom is the capacity to choose objectives consistent with one's values and within a larger social context (Logan and Stokes 2004, 38–39).

I also found the following description of the relationship of data and information that I accessed on Wikipedia on September 12, 2007 particularly illuminating:

> Even though information and data are often used interchangeably, they are actually very different. Data is a set of unrelated information, and as such is of no use until it is properly evaluated. Upon evaluation, once there is some significant relation between data, and they show some relevance, then they are converted into information. Now this same data can be used for different purposes. Thus, till the data convey some information, they are not useful.

I would interpret the signals transmitted between Shannon's sender and receiver as data. Consistent with MacKay and Bateson's position information makes a difference when it is contextualized and significant. Knowledge and wisdom represent higher order applications of information beyond the scope of this chapter. We will return, however, to the topic of knowledge and science in Chapter 8. The contextualization of data so that it has meaning and significance and hence operates as information is an emergent phenomenon. The communication of information cannot be explained solely in terms of the components of the Shannon system consisting of the sender, the receiver and the signal or message. It is a much more complex process than the simplified system that Shannon considered for the purposes of mathematizing and engineering the transmission of signals. First of all it entails the knowledge of the sender and the receiver, the intentions or objectives of the sender and the receiver in participating in the process and finally the effects of the channel of communication itself independent of its content as in McLuhan's (1964) observation that "the medium is the message". The knowledge and intention of the sender and the receiver as well as the effects of the channel all affect the meaning of the message that is transmitted by the signal in addition to its content.

The Meaning of Information in Biotic Systems

Biotic or instructional information, defined in POE as the constraints that allow an autonomous agent, i.e. a living organism, to convert free energy into work so that the living organism is able to propagate its organization through growth and replication, is intimately connected with meaning. "For Shannon the semantics or meaning of the message does not matter, whereas in biology the opposite is true. Biotic agents have purpose and hence meaning (Kauffman et al. 2007)." One can therefore argue that since the meaning of instructional information is propagating organization that we finally understand the meaning of life—the "meaning of life" is propagating organization. This remark is not meant to trivialize the great philosophical quest for the meaning of life from a human perspective but there is a sense in which the meaning of life including human life is indeed the propagation of organization. The purpose of life is the creation or propagation of more life.

In addition to the fact that Shannon information does not necessarily entail meaning whereas biotic or instructional information always entails meaning there is one other essential difference between the two. Shannon information is defined independent of the medium of its instantiation whereas biotic information is very much tied to its material instantiation in the nucleic acids and proteins of which it is composed. The independence of Shannon and cybernetic information from the medium of its instantiation is what gives rise to the notion of strong artificial intelligence and claims like those of Moravic, Minsky and to a certain extent Wiener that human intelligence and the human mind can some how be transferred to a silicon-based computer and does not require the wet computer of the human brain. Shannon and cybernetic information can be transferred from one material environment to another, from one computer to another or in the case of Shannon information from one telephone to another or from a computer to a hard copy of ink on paper. This is not the case with living organisms in the biosphere where information is stored in DNA, RNA and other structures of the organism such as their receptors for food/energy and danger/toxins.

One way of understanding our claim that biotic information contains meaning is to understand the relationship between life and agency, which arises as a emergent property of living systems. Kauffman (2008, 4) makes a distinction between "happenings" in the abiotic world and "doings" in the biosphere. Because of the fact that living systems have agency which manifests itself as their doing of things to insure the propagation of their organization. "Life, and with it agency, came naturally to exist in the universe. With agency came values, meaning, and doing, all of which are as real in the universe as particles in motion (ibid., x)." I would add that also with agency came purpose and with purpose information has meaning.

Shannon information whether on paper, a computer, a DVD or a telecommunication device, because it is symbolic, can slide from one medium or technology to another and not really change, McLuhan's "the medium is the message" aside. This is not true of living things. Identical genotypes can produce very different phenotypes depending on the physical and chemical environment in which they operate. Consider the fact that identical twins are not "identical". The reason identical twins are not "identical" is that the environment in which the biochemical interactions between biomolecules takes place alters the outcome.

The Materiality of Information in Biotic Systems

> *Information is information, not matter or energy. No materialism which does not admit this can survive at the present day.*
> *—Norbert Wiener (1948)*

> *Shannon's theory defines information as a probability function with no dimension, no materiality, and no necessary connection with meaning. It is a pattern not a presence.*
> *—Hayles (1999a, 18)*

Shannon information cannot be, nor was it meant to be, naively applied to complete living organisms, because the information in a biotic system like DNA is more than a pattern—it is also a presence. A receptor for food or toxins is not just a pattern—it is also a presence. A biological system is both an information pattern and a material object or more accurately information patterns instantiated in a material presence. Schrödinger (1944, 21) long ago before the discovery of DNA described this dual aspect of chromosomal material metaphorically. "The chromosome structures are at the same time instrumental in bringing about the development they foreshadow. They are law-code and executive power—or, to use another simile, they are architect's plan and builder's craft—in one." It is the dynamic of the interaction between the patterns of information and the material composition of the biotic agents that determines their behavior.

As previously discussed, the issue hinges on the degree to which one can regard a biotic agent as a fully physical computational system. It is clear that a biotic system cannot be described only by Shannon information for which the information is abstracted from its material instantiation and is independent of the medium. The same argument can be made for the inappropriateness of Kolmogorov information for biotic systems. Kolmogorov information, which is defined with respect to Turing machines, is another case where the information pattern is separated from its material instantiation. Biology is about material things not just mathematical patterns. As Kubie once warned at one of the Macy conferences, "we are constantly in danger of oversimplifying the problem so as to scale it down for mathematical treatment (Hayles 1999, 70)." As noted above the physical environment changes the meaning of the information embedded in the DNA of the genome.

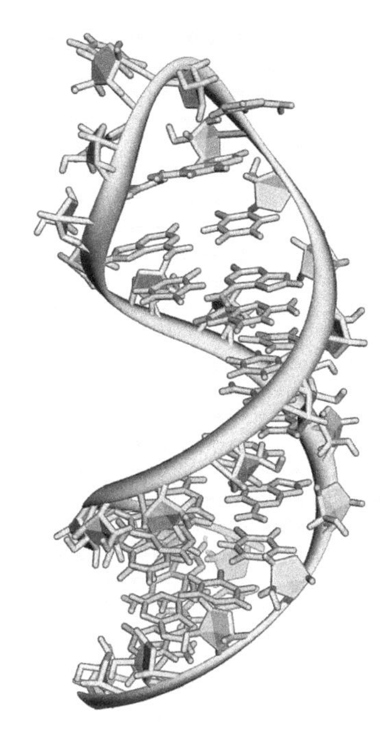

Hairpin loop, Pre-mRNA

Another way to distinguish the difference between biotic or instructional information and either Shannon or Kolmogorov information is that the latter are symbolic which is not the case for biotic or instructional information. The information coded in the chemical alphabet of biomolecules that make up living organisms acts through the chemical interactions of those biomolecules. "DNA is a molecule interacting with other molecules through a complex set of mechanisms. DNA is not just some text to be interpreted, and to regard it as such is an inaccurate simplification (Sarkar 1996, 860)." It is not the symbolic nature of DNA that gives rise to messenger RNA and it is not the symbolic nature of RNA that gives rise to proteins but rather the chemical properties of DNA that produce or catalyze the production of RNA and the chemical properties of RNA that produce or catalyze proteins and the chemical properties of proteins that carry out the protein's various functions such as:

1. serving as enzymes to catalyze biochemical reactions vital to metabolism,
2. providing structural or mechanical functions, such as building the cell's cytoskeleton,
3. playing a role in cell signaling, immune responses, cell adhesion and the cell cycle.

DNA, RNA and proteins are both the medium and the content, the message and the messenger. Not so for Shannon and Kolmogorov information where one can distinguish between the medium and the message, the message and the messenger. The message is the information, which operates independent of the medium in which it is instantiated, McLuhan aside. For biotic information, on the other hand, the medium and the message are the same—they cannot be separated. For biotic information the medium is the message in the McLuhan sense and it is also the content. For human symbolic information described by Shannon information, the information or content and the medium are quite separate. For biotic systems not only is the medium the message in the McLuhan sense that a medium has an effect independent of its content but the medium is also the content because it is the chemical properties of the medium that affect the organism. In fact the medium is the message because it is literally the content and the content of the message is unique to that medium and is instantiated in it and it cannot be transferred to another medium. To repeat it is not possible to transfer the content or the message of the medium to another medium. There is an isomorphism between the medium and its content. The medium is the content and hence also the message. The medium is both the message and the content for a biotic system because information in a biological system is not symbolic but rather chemical. It is for this reason that the notion of transferring the contents of the human brain to a computer is pure nonsense.

To conclude we have argued that information is not an invariant independent of the frame of reference in which it operates. In the biotic frame of reference information is always associated with meaning, which is not necessarily the case with Shannon or Kolmogorov information. In the biotic frame information cannot be separated from the medium of its instantiation as is the case in the Shannon and Kolmogorov reference frames. In other words the information in DNA, RNA and proteins are embodied. They differ from human symbolic information, which can be disembodied and moved from one medium to another. Each generation makes a god of their latest technological or scientific achievement or breakthrough. For the Hebrews it was the written word and the law "written with the finger of God". For the Greeks it was their deductive logic and rational thought disembodied from practical experience and empirical evidence of the physical world. For the Enlightenment it was Newtonian mechanics and God, the clock maker, where things were explained in terms of mechanical models. In the Information Age the god is disembodied information, information without context where everything is explained in terms of the transfer of information, and some times it is information without meaning.

Organization as Information

What is the relationship of organization and information? What we discovered in POE was that the autocatalysis of biomolecules led to the organization of a biological living organism whose organization of constraints allowed it to convert free energy into work that sustained growth and permitted replication. We identified the constraints as instructional or biotic information, which loops back into the organization of the organism. This model of information holds for biotic systems where collective autocatalysis is the organization and the components are the individual biomolecules.

The argument seems circular only because a living organism represents a self-organizing system. This is still another way that biotic information differs from Shannon information which is defined independent of meaning or organization. In fact organized information has less Shannon information because it does not reduce as much uncertainty as disorganized information.
It is also the case as we mention above that this model provides a mechanism for the creation of information which in not the case with the Shannon model of information.

I believe that Hayles (1999a, 11) has come to a similar conclusion regarding the relationship of information and organization when she wrote about the paradigm of autopoiesis or self-organization:

> Information does not exist in this paradigm or that it has sunk so deeply into the system as to become indistinguishable from the organizational properties defining the system as such.

It is the latter half of her statement that is congruent with our notion that the set of constraints or organization that give rise to an autonomous self-organizing system is a form of information.

Wiener like Shannon related information to entropy but, unlike Shannon, Wiener (1948, 18) saw a connection between organization and information, "The notion of the amount of information attaches itself very naturally to a classical notion in statistical mechanics: that of entropy. Just as the amount of information in a system is a measure of its degree of organisation, so the entropy of a system is a measure of its degree of disorganisation."

We (Kauffman et al. 2007) made a similar claim in POE when we asserted that the constraints that allow the propagation of organization in a living organism represents the information content of that organism. In other words the propagating organization of a living organism is its information content. Our position in a certain sense recapitulates similar sentiments expressed by Norbert Wiener (1954, 96) when he wrote "We are not stuff that abides but patterns that perpetuate themselves."

However where I differ from Weiner is that while we are patterns that abide I also believe that we are patterns that are uniquely instantiated in flesh. I therefore believe that human intelligence cannot be transferred from a human brain onto a silicon-based computer as is claimed by some advocates of strong AI. The point that I would make is that the pattern cannot be separated from the medium in which it is instantiated as was argued above. The medium of flesh and its organization are what is critical. It is the pattern instantiated in the flesh and not just the pattern by itself that makes life. The information in a biological system is not symbolic but rather chemical. As we have already asserted the medium of the flesh is both the message and the content of a biotic system. We will return to the question of the relationship of information and organization once we have introduced some ideas about the origin of language and culture and their relationship to information. But before addressing this issue let us continue our discussion of whether or not living organisms are information or flesh.

Who Are We? What Are We, Information or Flesh?

Information in the form of words or language is symbolic. The word cat is a symbol that represents a class of living breathing creatures made of flesh. An actual cat is not a symbol of something else but an organization of organic chemicals that can propagate its organization through its metabolism and its ability to replicate.

The organic chemicals of which we are composed are continually replaced so that after seven years there is a completely new set of molecules. So we are not flesh or a particular set of molecules but the organization of the molecules of which we are composed or more accurately we are a process and not a thing that can be duplicated.

One cannot make a replica of a person. Even twins that originated from the same fertilized egg are never exactly the same. But a text can be replicated or duplicated exactly. A text can also be transmitted and reformatted from one medium to another, for example from a computer file to a text printed on paper or from a live performance to a podcast.

I believe that the proponents of strong artificial intelligence (AI) and strong artificial life (AL) make the mistake of considering intelligence or life as merely reified information. They do not take into account that it is the interaction or organization of flesh-based matter that makes intelligence and life. The pattern of that interaction or organization that we identify as information cannot be abstracted away from the physical medium in which it is instantiated and remain unchanged or, even more importantly, continue as the process that gave rise to that intelligence or life in the first place.

A feature of both intelligence and life is that it is autonomous. A living organism is an autonomous agent that has the capacity to exploit free energy from its environment and use that energy in the form of work to carry out its metabolism, to replicate and to make use of its intelligence. The proponents of strong AI and AL overlook this important factor when they claim that intelligence and life is nothing more than information or a pattern that is independent of its physical instantiation. At best artificial life forms may be regarded as obligate symbionts with humans but not as independent living organisms as they are not autonomous.

We attempted to answer the question who are we, what are we: information or flesh? Our conclusion is that we are both but in order to more fully address the question we need to deal with the issue of language. In the next chapter we will examine the role of language in defining who and what we are. We will discover that language is a critical element of determining who and what we are because of the way that language extends the brain into the human mind and creates the conditions for the emergence of culture another unique feature of humankind that defines us.

Acknowledgement: This chapter draws heavily on two sources other than my earlier work, namely the paper Propagating Organization: An Enquiry (Kauffman et al. 2007) that I co-authored and the book *How We Became Posthuman* (Hayles 1999a). In a certain sense this chapter is a remix of these two sources with help from the cited references.

Chapter Three

Information and Language and their Interrelationship

In the last chapter we made a distinction between biotic information like that contained in DNA and human symbolic information which is instantiated in language. In this chapter we examine the relationship of symbolic information with language, the human mind, and thought.

The connection of information to the mind can be traced back to the original definition of the word inform which meant to give a form to the mind.

In addition to this connection it is almost impossible to think of thought and information that is not connected to one form of language or another. Language is the medium through which symbolic information is formulated and communicated with the exception of the visual arts and music (to be treated in Chapter 10) and as such deserves our special attention. As we will discover language is both a medium of communication and an informatics tool, which I have formulated in terms of the equation: language = communications plus informatics (Logan 1995 & 2004b). We will therefore describe the origin of language in this chapter and its connection to thought and information (Logan 2007). We will also describe the evolution of language from its very first form as speech into writing, mathematics, science, computing and the Internet (Logan 2004b). We examine the different forms of the notation of language and

Marshall McLuhan

information and show how they impact the way in which we think, organize our institutions, develop our tools and interact with each other. In particular we will study the way in which the phonetic alphabet and the place number system have led to the digital information explosion of the past 50 years (Logan 2004a & McLuhan and Logan 1977). For readers not familiar with my earlier work (Logan 2004a &b and 2007) I have summarized them in this chapter. In fact the headings of the following sections are the titles of those books.

The Extended Mind Model of the Emergence of Language and the Human Mind

> *All media are active metaphors in their power to translate experience into new forms. The spoken word was the first technology by which man was able to let go of his environment in order to grasp it in a new way.*
> *—Marshall McLuhan, Understanding Media*

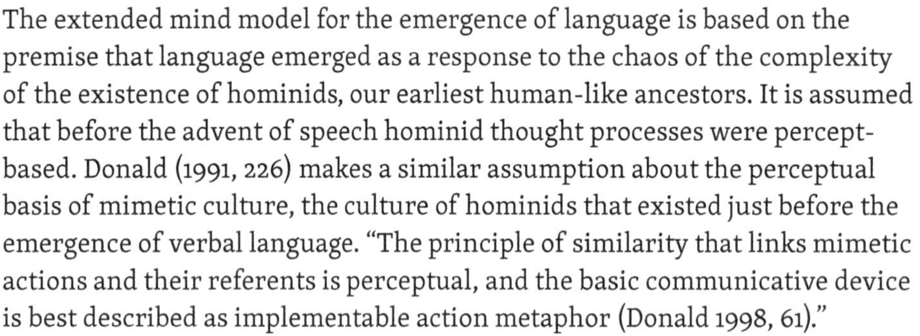

The extended mind model for the emergence of language is based on the premise that language emerged as a response to the chaos of the complexity of the existence of hominids, our earliest human-like ancestors. It is assumed that before the advent of speech hominid thought processes were percept-based. Donald (1991, 226) makes a similar assumption about the perceptual basis of mimetic culture, the culture of hominids that existed just before the emergence of verbal language. "The principle of similarity that links mimetic actions and their referents is perceptual, and the basic communicative device is best described as implementable action metaphor (Donald 1998, 61)."

Hominids that emerged in the savannas of Africa were an easy target for various predators. To defend themselves from this threat and to increase their food supply they acquired the new skills of tool making, the control of fire, and coordinated group foraging and hunting as has been described by Merlin Donald (1991) in his book *The Making of the Modern Mind*. To pursue these activities a more complex form of social organization emerged among hominids including non-verbal mimetic communication, further increasing the complexity of their existence. This complexity could be handled at first through more sophisticated percept-based responses that characterized mimetic culture (ibid.). At some point, however, the complexity of hominid existence became too great. Percept-based thought alone did not provide sufficient abstraction to deal with this increased complexity. The hominid brain could no longer cope with the richness of its life based solely on its perceptual sensorium. In the information overload and chaos that ensued, I believe, a new abstract level of order emerged in the form of verbal language and conceptual thinking.

In fact our first words were our first concepts and they allowed for a more abstract form of thought. Words did not give rise to concepts nor did concepts give rise to words, rather human language and conceptualization emerged at exactly the same point in time creating the conditions for their mutual emergence. This transition or bifurcation from the concrete percept-based thinking of pre-lingual hominids to conceptual-based spoken language and thinking is an example of punctuated equilibrium. I believe this transition was the defining moment for the emergence of the fully human species Homo sapiens. This discontinuous transition illustrates Prigogine's theory of far from equilibrium processes and the notion that a new level of order can suddenly emerge as a bifurcation from a chaotic non-linear dynamic system (Prigogine and Stengers 1984 & Prigogine 1997).

Words represent concepts and concepts are represented by words. Each word serves as a metaphor and a strange attractor uniting all of the pre-existing percepts associated with that word in terms of a single word and, hence, a single concept. All of one's experiences and perceptions of water, the water we drink, bathe with, cook with, swim in, that falls as rain, that melts from snow, were all captured with a single word, water, which also represents the simple concept of water. Not all the words of our language emerged as generalizations of percepts but once language emerged it allowed an abstract level of thought that led to the creation of new function words that served grammatical purposes and gave syntactic structure to language. Syntactical structures are also concepts. They are concepts that encompass relationships between words just as words are concepts that encompass relationships between percepts.

Percepts are the direct impressions of the external world that we apprehend with our senses. Concepts, on the other hand, are abstract ideas that result from the generalization of particular examples. Concepts allow one to deal with things that are remote in both the space and time dimensions and they make it possible to model the external world and plan. Humans are the only animals capable of language and planning. These ideas parallel the work of Lev Vygotsky (1962) in his seminal work *Language and Thought*.

We attributed the emergence of language due to the complexity of hominid life due to the control of fire, tool making, complex social structures, coordinated large scale hunting and gathering and pre-lingual mimetic communication consisting of hand signals, gestures, body language and non-verbal vocalizations. These activities actually served as a cognitive laboratory for the development of verbal language. According to Christiansen (1994) toolmaking entailed sequential learning and processing, which could have also served as a pre-adaptation for speech.

Complex social structures led to social intelligence and the need to cooperate and share information and hence also served as another pre-adaptation for speech. The information overload of interacting with many people and carrying out more sophisticated activities led to the need for better communications to better co-ordinate social transactions and co-operative activities such as the sharing of fire, the maintenance of the hearth, food sharing, and large scale co-ordinated hunting and foraging. From the chaos of this complexity emerged the preverbal protolanguage of mimetic communication, which according to Donald (1998, 61) "establishes the fundamentals of intentional expression in hominids, without which language would not have had an opportunity to evolve such a sophisticated, high-speed communication system as modern language unless there was already a simpler slower one in place." It was in the context of this early form of hominid communication that the skills of generativity, representation and communication developed.

Toolmaking, social interaction and mimetic communication gave rise to more than just spoken language and conceptual thinking. Transformed by the verbal language and conceptual thinking they gave rise to, they also served as the prototypes for three fundamental activities that form the core of modern human society, namely technology which emerged from tool making; commerce which emerged from social organization and intelligence; and the art forms which emerged from mimetic communication. "There is a vestigial mimetic culture embedded within our modern culture and a mimetic mind embedded within the overall architecture of the modern human mind (Donald, 1991, 162)."

The Extended Mind

Not only did language transform toolmaking, social interaction and mimetic communication respectively into technology, commerce and the fine arts but it also transformed or extended the brain into the human mind. For many psychologists the brain and the mind are synonymous, just two different words to describe the same phenomena, one derived from biology, the other from philosophy. Others define the mind as the seat of consciousness, thought, feeling and will while those processes of which we are not conscious are not activities of our mind but functions of our brain. There is no objective way to resolve these two different points of view but I believe that a useful distinction can be made between the mind and the brain based on our dynamic systems model of language as the bifurcation from concrete percept-based thought to abstract concept-based thought. I, therefore, assume that the mind came into being with the advent of verbal language and, hence, conceptual thought.

Verbal language extended the effectiveness of the human brain and created the mind. Language is a tool and as all tools are extensions of the body according to McLuhan (1964) it follows that language extended the brain into the mind or what I have termed the extended mind. I have expressed this idea in terms of the equation: mind = brain + language.

This hypothesis is itself an literal extension of the ideas of McLuhan's (1964):

> It is the extension of man in speech that enables the intellect to detach itself from the vastly wider reality. Without language, Bergson suggests, human intelligence would have remained totally involved in the objects of its attention. Language does for intelligence what the wheel does for the feet and the body. It enables them to move from thing to thing with the greatest ease and speed and ever less involvement. Language extends and amplifies man.

The human mind is the verbal extension of the brain, a bifurcation of the brain, which vestigially retains the perceptual features of the hominid brain while at the same time becoming capable of abstract conceptual thought. Andy Clark has also independently developed the notion of "the extended mind" (Clark 1997; Clark and Chalmers 2003).

The emergence of syntactilized language also represents, for me, the final bifurcation of hominids from the archaic form of Homo sapiens into the full-fledged human species, Homo sapiens. Crow (2002, 93) reaches a similar conclusion, "The parsimonious conclusion ... is that the origin of language coincided with the transition to modern Homo sapiens dated to somewhere between 100,00 and 150,000 years ago."

Humans are the only species to have developed verbal language and also to have experienced mind. Our ancestors, the earlier forms of hominids, experienced thought but their thought patterns were percept-based and their brains functioned as percept processing engines operating without the benefit of the abstract concepts, which only words can create and language can process. It follows that animals have brains but no minds and that the gap between humans and animals is that only humans possess verbal language and a mind.

In summary, the emergence of verbal language represents three separate bifurcations:

1. the bifurcation from percepts to concepts,
2. the bifurcation from brain to mind, and
3. the bifurcation from archaic Homo sapiens to modern human beings, Homo sapiens.

These three bifurcations are not necessarily simultaneous. Bickerton claims (1990, 1995) that protolanguage in which the first words were used symbolically emerged with Homo erectus which means the first bifurcation can be dated to approximately 2 million years ago. The second and third transitions, on the other hand, can be dated to the emergence of fully syntactilized language, which occurred only 100±50 thousand years ago and seems to be correlated with the explosion of human culture and technological progress of that time period (Bickerton 1995, 65).

Language as a Living Organism

The approach we have developed to understanding the origin of language is based on the notion that the skills acquired from toolmaking, social interaction and mimetic communication were the pre-adaptations for the emergence of human language. This differs markedly from the explanation provided by Noam Chomsky who pioneered the concept that all the world's languages share a Universal Grammar (UG). Chomsky's explanation for this is that because of a genetic mutation we are all hard wired with UG. In addition to this he postulated that we are also hard wired with a language acquisition device (LAD), which explains why young children are able to learn their parent's language effortlessly and automatically despite the poverty of stimulus in their language learning. They do not need to be explicitly taught the rules of UG but they seem to be able to use them without instruction. In the Extended Mind model it is assumed that it is by mimesis that youngsters pickup their parent's language and that the universality of the world's languages is due to the universality of human cognitive abilities and the fact that toolmaking, social interaction and mimetic communication served as the pre-adaptations for the emergence of human language.

Morten Christiansen's (1994) proposed another alternative to Chomsky's hard wired hypothesis consistent with the Extended Mind model when he suggested that language could be considered as an organism that evolved to be easily learned especially by children despite the poverty of stimulus problem. In a later paper with his colleagues he wrote:

> Language exists only because humans can learn, produce, and process them. Without humans there would be no language. It therefore makes sense to construe languages as organisms that have had to adapt themselves through natural selection to fit a particular ecological niche: the human brain. In order for languages to 'survive', they must adapt to the properties of the human learning and processing mechanisms. This is not to say that having a language does not confer selective advantages onto humans. It seems clear that humans with superior language abilities are likely to have a selective advantage over other humans (and other organisms) with lesser communicative powers. This is an uncontroversial point, forming the basic premise of many of the adaptationist theories of language evolution. However, what is often not appreciated is that the selection forces working on language to fit humans

> are significantly stronger than the selection pressures on humans to be able to use language. In the case of the former, a language can only survive if it is learnable and processable by humans. On the other hand, adaptation toward language use is merely one out of many selective pressures working on humans (such as, for example, being able to avoid predators and find food). Whereas humans can survive without language, the opposite is not the case. Thus, language is more likely to have adapted itself to its human hosts than the other way around. Languages that are hard for humans to learn simply die out, or more likely, do not come into existence at all. (Christiansen, Dale, Ellefson and Conway 2001, 144–45)

This hypothesis at once explains why grammars are universal and are easily learned by children with out the need of a hard-wired LAD.

Stuart Kauffman's (1995, 49) notion of the reproduction of living organisms by autocatalysis can be applied to language operating as an organism. "A living organism is a system of chemicals that has the capacity to catalyze its own reproduction." Let us use Kauffman's definition and apply it in a generalized form to language operating as an organism. We are justified to regard language as a living organism because it is a system of words and grammatical structures that has the capacity to catalyze its own reproduction. If we consider each person's individual use of language as an organism then we may regard language reproducing itself each time a child acquires his or her parents' language. With this definition we not only meet Kauffman's criteria of that an organism catalyzes its own reproduction but we are able to consider the evolution of this organism using Darwin's simple one line definition of evolution, namely, "descent with modification." For Darwin descent meant reproduction. By considering the language of each individual in the society as an organism we can speak of a language reproducing itself. In this case the inheritance or descent is not by diploidy but the polyploidy of parents, siblings, peers, teachers, relatives and society in general. One can now apply the concept of natural selection to the language organism of each individual in a society.

The language of the society as a whole is not an organism because it cannot reproduce itself but we can think of it as a species of organisms consisting of the languages of all the individuals of the society. De Saussure called the language of society "langue," and the language practiced by individuals "parole". With our definitions "langue" is the species and "parole" is the organism. Just as conspecifics of a biological species are able to reproduce among themselves the conspecifics of a linguistic species are able to communicate with each other. English and French are linguistics species. The linguistic competence of each individual represents an organism with its own unique language, which can communicate only with members of its linguistic species.

The ontogeny of language acquisition is both biological and cultural in that the child must be born with the genetic apparatus for speech and at the same time be exposed to a language. A linguistic species like English or French does not belong to an individual but to a community or a culture. Dawkins (1989, 192) has argued that Darwinian principles apply with the same validity to cultural replicators or memes as they do to biological replicators, namely genes. "Just as genes propagate themselves in the gene pool by leaping from body to body via sperm or eggs, so memes propagate themselves in the meme pool by leaping from brain to brain via a process which, in the broad sense, can be called imitation."

Robert Worden (2000) in an attempt to understand how language changes suggested a language as langue can be considered as an ecology populated by words each of which are memes that interact with each other. Intrigued by Worden's idea I extended it by proposing that every word and every grammatical construction of spoken and written language as well as every semantic element and syntactical structure of math, science, computing and the Internet are memes just like the other elements of culture. If a new word is used to refer to a new experience or a new syntactical structure is used to refer to a new relationship and it is copied by a listener and replicated that word or syntactical structure fits Dawkin's definition of a meme. Words and syntactical structures evolve and compete. They are adaptations and they contain vestigial structures. They are living entities that are part of a living language and they like biological systems evolve and compete. They differ in that they are information without extension rather than a living organism made up of atoms that occupy physical space. One can think of them as living systems of information or organization that propagate themselves in the same way that Dawkins (1996, 81) does. "Language seems to 'evolve' by non-genetic means, and at a rate which is orders of magnitude faster than genetic evolution."

The Sixth Language and the Evolution of Notated Language

Every particular notation stresses some particular point of view.
—Ludwig Wittgenstein

The memetic thinking of Dawkins and Worden intersects with my earlier work on the evolution of language in a book entitled The Sixth Language (Logan 2004b) in which I demonstrated that speech is part of an evolutionary chain of languages which also includes writing, mathematics, science, computing and the Internet. Each of these languages, as we will see, acts as cultural replicators that propagate their organization through their memes. And like biological organisms they are living cultural entities that evolve, compete and pass on some form of information.

Linguists traditionally define language strictly in terms of communication. One example is Edward Sapir's (1921) definition of language is as "a purely human and non-instinctive method of communicating ideas, emotions, and desires by means of a system of voluntarily produced symbols." Communication is not the only function of language. Language also plays a key role in the formulation of information, including its processing, storage, retrieval, and organization. Language is a tool for developing new concepts and ideas as pointed out by Vygotsky (1962) and others. Writing, mathematics, science, computing and the Internet permit the development of ideas that could never arise through the use of speech alone. We must therefore consider these other modes as distinct, albeit related, languages.

Generalizing and extending Sapir's definition, I define language as "a purely human and non-instinctive method of communicating ideas, emotions, and desires as well as a systems formulating, processing, storing, retrieving, and organizing information by means of a system of voluntarily produced symbols." Speech therefore is not the sole form of language. We claim that speech and writing, for instance, are two distinct but related forms or modes of language. This position differs from the beliefs of traditional linguists who consider speech as the primary form of language and writing as merely a system for transcribing or recording it. Ferdinand de Saussure (1967), one of the founders of the field of linguistics, wrote: "Language and writing are two different systems of signs; the latter only exists for the purpose of representing the former.... The subject matter of linguistics is not the connection between the written and spoken word, but only the latter, the spoken word is its subject." Leonard Bloomfield (1933, 21) wrote: "Writing is not language, but merely a way of recording language by means of visible marks." These linguists did not understand the cognitive implications of language like Stubbs and Chafe. "Writing is not merely a record.... I know from personal experience that formulating ideas in written language changes those ideas and produces new ones (Stubbs 1982)." "Language is used in a variety of ways, each of which affects the shape that language takes. Since the 1970s, ever-increasing attention has been paid to differences between spoken language and written language, and it has become clear that each of these two broad categories allows for diverse uses and forms (Chafe 1998, 96)."

Although written language is derived from spoken language, it is useful to regard them as two separate language modes because they process information so differently. A similar argument can also be made that mathematics, science, computing, and the Internet should be regarded as separate language modes. Each of these five modes of language has unique strategies for communicating, storing, retrieving, organizing, and processing of information. I have therefore extended the notion of those linguists who consider speech and writing as separate modes of language to claim that speech, writing, mathematics, science,

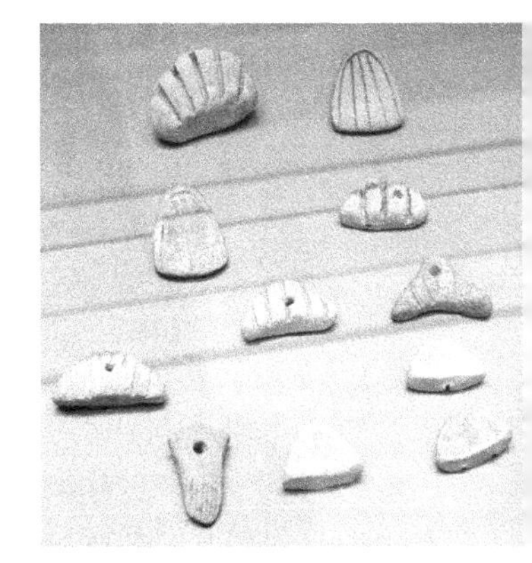

Clay tokens

computing, and the Internet are six separate modes of language, which are distinct but interdependent. They form an evolutionary sequence in which the later modes are derived from and incorporate elements of the earlier modes of language. They form a nested set of languages in which the later forms contain all of the elements of the earlier forms.

Speech, the first form of human language, is the basis of all other linguistic modes of communication and information processing. We can define spoken language as the sum of information uttered by human speakers. Written language, which is derived from speech, is defined as the sum of information, which has been notated with visual signs. It differs from speech in that it involves a permanent record, whereas speech disappears immediately after it is uttered. We shall distinguish five different modes or forms of notated language: writing (or text), mathematics, abstract science, computing and the Internet.

Writing and mathematical notations were the first forms of written language; both grew out of the system of recording payments of agricultural tributes using clay accounting tokens in Sumer just over 5,000 years ago. They emerged to deal with the information overload and the limits on human memorization that the accountants had to deal with in keeping track of the tributes in the form of agricultural commodities paid by farmers to the priesthood. These commodities were then distributed to the irrigation workers whose efforts allowed the water of the Tigris-Euphrates river system to flow into the farmers' fields. Because writing and mathematical calculations had to be taught the first formal schools arose in Sumer. The teachers began to collect information for their lesson plans and from this activity scholarship and eventually science emerged. Science emerged as organized knowledge to deal with the information overload created by teacher/scholars. The methods and findings of science are expressed in the languages of writing and mathematics, but science may be regarded as a separate form of language because it has a unique way of systematically processing, storing, retrieving, and organizing information, which is quite different from either written text or mathematics. A little more than sixty years ago, the next system for processing information emerged from science and mathematics in the form of computing, with its own unique cybernetically based and automated methods for processing and organizing information. Computing arose to deal with the information overload of science and science based technology. Finally, the latest form of language emerged from computing and telecommunications in the form of the Internet and the World Wide Web. The Internet emerged as a way of dealing with the information overload generated by computing and the need to store and transmit all of that information.

Whether these six modes of information processing and communication should be regarded as separate languages or whether they are merely six different aspects of the human capacity for language are questions we will not address. For the purposes of our analysis, we will consider speech, writing, mathematics, science, computing, and the Internet as six distinct modes of language, which form an evolutionary chain of development. What these modes of language share is a distinct communications and informatics methodology. Each provides a unique framework for viewing the world.

Clay envelope

The Semantics and Syntax of the Six Modes of Language

The claim made here that writing, mathematics, science, computing, and the Internet are distinctive modes of human language is based on the notion that a language is defined by both its informatics and communication capacity. These five modes of language may be regarded as languages in their own right because of their unique semantics and syntax. The traditional linguists Paivio and Begg (1981, 25) argued, "semantics and syntax—meaning and grammatical patterning—are the indispensable core attributes of any human language." Semantics relates a signal to its meaning whereas syntax is concerned with the structure or relationship among linguistic signals.

New modes of language evolved to represent increasingly more complex phenomena, and hence, according to Ashby's law of requisite variety (Ashby 1957), they required a richer vocabulary and more complex structures to function. We therefore expect the semantics and the syntax of the new forms of language to retain the older structures and add their own new unique elements to those structures.

Writing, mathematics, science, computing, and the Internet are distinct modes of human language, which differ from speech, because they have distinct semantic and syntactical features above and beyond those of speech. The semantics of the written word are quite similar to those of the spoken word, but there are large numbers of exceptions. Many constructions, which are acceptable in oral language, are not valid in prose and vice versa. One's written vocabulary is considerably greater than one's oral vocabulary. We often use words in our written communication that we would never use in our oral discourse. For example there are many more abstract words in written prose. A comparison of the lexicon of Homer (as found in the transcriptions of his orally composed poems) and that of the ancient Greek philosophers and playwrights of the fifth and fourth centuries BCE reveals the development of a new written vocabulary. The new lexicon of written words is rich in abstract terminology appearing in the language for the first time, and old words take on additional new abstract meanings (Havelock 1963).

It is syntactically where written language begins to diverge more radically from speech. Writing encourages a formal structuring of language consisting of sentences, paragraphs, and chapters largely absent in spoken language.

Akkadian clay tablet

Analysis of a transcription of oral discourse reveals that spoken language is often grammatically incorrect. The term grammar betrays its association with writing through its etymology. The Greek term for "letter" (as in letters of the alphabet) is gramma. Grammar was not formalized before writing, just as there was no such thing as spelling before writing, and no uniform spelling before the printing press.

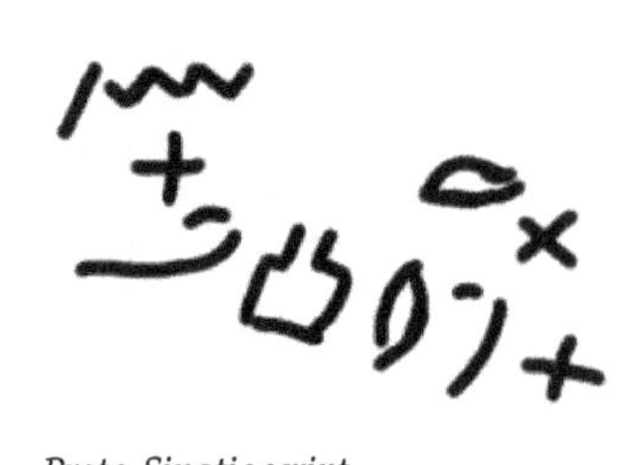

Proto-Sinatic script

In the language of mathematics, the semantic domain or lexicon consists primarily of precisely defined notations for numbers such as 0, 1, ½, 0.4, and the square root of 2; mathematical operations such as +, -, ×, and ÷; and logical relationships such as >, <, and =. The other semantic elements unique to mathematics are its definitions and axioms, such as those found in geometry, number theory, and other logical systems. The language of mathematics differs from natural language such as spoken English in that the semantic relationship between the sign such as a numeral—and the concept being represented, an abstract number, is totally unambiguous. The precision of the semantic conventions of mathematics also extends to the syntactical domain. The basic syntax of the language of mathematics is that of logic. Mathematical syntax, unlike that of spoken or written language, is totally unambiguous. The rules of grammar that govern speech and writing are subject to conflicting interpretations while those of mathematics are not. The language of mathematics also introduces unique syntactical structures not found in natural languages, such as proofs, theorems, and lemmas.

The language of science includes many of the semantic elements of speech, writing, and mathematics, but it also introduces new semantic elements unique unto itself. These include quantitative concepts like velocity, mass, and force; qualitative concepts like organic/inorganic, solid/liquid/gas; and theoretical concepts like inertia, entropy, valence, and natural selection. Like the language of mathematics, the semantics of science is characterized by precise and unambiguous definitions even though much of the terminology that is employed corresponds to words that appear in everyday spoken language. In spoken English, mass can refer to either volume or weight. In physics mass is precisely defined in terms of its gravitational and inertial properties.

The syntax of science includes the structure of speech, writing, and mathematics. Science also introduces its own syntactical elements, however. The three most important elements, the ones, which in a sense define the nature of science, are: (1) the scientific method; (2) the classification of information or data (taxonomy); and (3) the organization of knowledge such as the grouping of scientific laws to form a scientific theory. The centrality of the classificatory and organizational structures is due to the fact that science is defined as organized knowledge. The scientific method, with its elements of observation, generalization, hypothesizing, experimental testing, and verification, is the key element, which defines the character of science. It is the scientific method that qualifies science as a distinctive language rather than a carefully organized scholarly activity like history, which also makes use of organizational principles and other modes of language, namely, speech, writing, and mathematics.

The language of computing includes all of the semantic and syntactical elements of the earlier four modes of language. It also possesses its own semantic and syntactical elements by virtue of the activities of both its programmers and its end users. The semantics of the programming languages and end-user software programs specify computer inputs and outputs. The syntactical structures of programming languages and end users' software formalize the procedures for transforming inputs into outputs. These syntactical structures are basically unambiguous algorithms for ensuring the accuracy and the reliability of a computer's output. The syntactical structures that arise in a programming language or a relational database differ from the other language modes so that the user can take advantage of the computer's rapid information-processing speeds.

Although the Internet and the World Wide Web incorporate all of the semantic and syntactical elements of computing they also include their own unique elements in both categories. Perhaps we should clarify the relationship between the Web and the Net. The World Wide Web is one of the many different elements of the Internet, which include its email facilities, listservs, chat rooms, ftp facilities, Telnet facilities, Web pages, Web sites, intranets, extranets, portal sites and e-commerce sites. Each of these facilities represents the semantic elements of the sixth language of the Internet.

The Internet has a number of unique syntactical elements. One of the unique syntactical elements of the sixth language is hypertext, which makes it possible to link all of Web sites and Web pages in cyberspace to form one huge global document. Another unique syntactical element is the Internet Protocol, which allows all of the computers connected to the Internet to form one huge Global Network and makes the Web, ftp and telnet all possible. McLuhan's prediction of a Global Village has been realized. Still another unique syntactical element of the Internet are the search engines, which increase access to knowledge and information and hence provide an extra level of communication that the other forms of verbal language cannot match. The search engine also facilitates people finding each other and hence contributes to the creation of a global knowledge community.

The Alphabet Effect

The work we have reviewed here on the origin and evolution of language is based on the notion that language is both a medium of communication and an informatics tool. This idea grew out of the study that McLuhan and I made of the impact of phonetic writing and the phonetic alphabet on the development of Western culture where we demonstrated that the alphabet while primarily a medium of communications had an enormous impact on the way information was formulated and organized. What gave rise to this study was an attempt to understand why abstract science had begun in the West and not in China where so much of the world's technology originated such as animal harnesses, iron and steel metallurgy, gunpowder, the drive belt, the chain drive, the standard method of converting rotary to rectilinear motion, and the segmental arch bridge. To this must be added irrigation systems, paper, ink, printing, movable type, metal-barrel cannons, rockets, porcelain, silk, magnetism, the magnetic compass, stirrups, the wheelbarrow, Cardan suspension, deep drilling, the Pascal triangle, pound-locks on canals, fore-and-aft sailing, watertight compartments, the sternpost rudder, the paddle-wheel boat, quantitative cartography, immunization techniques (variolation), astronomical observations of novae and supernovae, seismographs, acoustics, and the systematic exploration of the chemical and pharmaceutical properties of a great variety of substances. Having carefully documented through years of historical research the contribution of Chinese science and its influence on the West, Needham (1979, 11) posed the following question: "Why, then, did modern science, as opposed to ancient and medieval science, develop only in the Western world?" By modern science he meant abstract theoretical science based on experimentation and empirical observation, which began in Europe during the Renaissance.

I proposed that since monotheism and codified law were unique to the West and that together they give rise to a notion of universal law that this might explain the Needham paradox. I shared these thoughts with Marshall McLuhan who immediately pointed out that the alphabet which served as a model for analysis, classification, coding and decoding was also unique to the West. We (McLuhan & Logan 1977) combined our ideas and developed the hypothesis that the phonetic alphabet, codified law, monotheism, abstract science and deductive logic were ideas unique to the West and while they were not causally linked, they were mutually self-supporting or autocatalytic. I carried away from this work on the alphabet effect (Logan 2004a) the understanding that the way in which a language was notated could effect the way its users think and develop concepts.

The Mesopotamian phonetic syllabary inspired the organization of social mores into forms of codified law, the most famous of which is the Hammurabic code. The impact on the Hebrews of the alphabet, which they borrowed from the Midianites was immediate and dramatic. In addition to bringing writing to the Hebrew children, Moses also brought them codified law, or the Ten Commandments, as well as a more abstract and monotheistic concept of God.

The introduction of the phonetic alphabet into Greek society had an equally dramatic effect on that culture. Alphabetic writing promotes analysis because each word must be analyzed into its basic phonemes in order to be

transcribed. When spoken language is transcribed, phonemes or sounds are coded with meaningless visual signs, the letters of the alphabet. And when written text
is read those visual signs are decoded back into sounds. The alphabet is also a classification tool which allows a perfect ordering through alphabetization of all of the spoken words of any language transcribed with an alphabetic writing system. The analysis, coding, decoding and classification that the phonetic alphabet promotes are the basic ingredients of abstract science and deductive logic.

Parminides

Within 500 years of the transmission of the alphabet from the Phoenicians, the Greeks developed the main intellectual concepts, which have formed the foundation of Western civilization. They created, for the first time in history, abstract science, formal logic, axiomatic geometry, rational philosophy, representational art, and individualism. While not suggesting a causal connection between these developments and the alphabet, I believe that the alphabet, by serving as a paradigm for classification, analysis, and codification, created the conditions that made these new ideas possible.

Another fallout from alphabetic writing was the invention of zero and the place number system. The place number system and the concept of zero were inventions of Hindu mathematicians as early as 200 BCE. The Hindu writing system at the time was alphabetic, as was their number system. Once the Hindu mathematicians developed the notion of zero, or sunya, as they called it, they quickly devised a place number system.

Sunya means "leave a place" in Sanskrit and indicates that the zero or sunya concept arose from recording abacus calculations. If the results of an abacus calculation was 503, this could not be written as "5" "3" (where "5" and "3" were the alphabetic symbols for the numerals 5 and 3) because "5" "3" would have been read or interpreted as 53 or 530 and not 503, but if instead the result was written as "5" "leave a place" "3", the number being designated would be interpreted properly as 5 hundreds, no tens, and 3 units and hence 503. "Leave a place" or sunya soon evolved into the abstract number of zero, and was represented at first by a dot . and later by today's current symbol, 0.

The Arabs used the Hindu system and transmitted it to Europe, where it arrived in the fifteenth century. The Arabs had translated sunya or "leave a place" into the Arabic sifr or cipher, the name we still use for zero as well as the name for the whole place number system itself. Our present-day term zero derives from the shortened version of the Latin term for cipher, zepharino. The place number system brought with it many advances in mathematics, including simple algorithms for arithmetic, negative numbers, algebra, the concept of the infinite and the infinitesimal, and hence, calculus. Cipher also means a secret code because at first the Vatican banned the use of Arabic numerals but Italian merchants used them secretly.

One of the mysteries associated with the invention of the place number system is why the Greeks, the inventors of vowels, who made such great advances in geometry and logic, did not discover zero. The explanation lies in the Greeks' overly strict adherence to logic, which led Parmenides to the conclusion that non-being (and hence, nothing) could not "be" because it was a logical contradiction. The Hindus, on the other hand, had no such inhibition about non-being. In fact, they were positively inclined to the concept of non-being since it constituted their notion of nirvana (Logan 2004a; Logan 1979, 16).

The most dramatic communication revolution to follow the introduction of the alphabet was the Gutenberg printing press. The revolutionary changes brought about by this technology have been documented by McLuhan in The Gutenberg Galaxy, a seminal work in which he shows the impact of print on such major cultural transformations as the rise of science, the Reformation, the Enlightenment, the rise of nationalism, and the Industrial Revolution. "The invention of typography confirmed and extended the new visual stress of applied knowledge providing the first uniformly repeatable 'commodity,' the first assembly-line, and the first mass production" (McLuhan 1962, 153).

Human Language, Culture, Technology, Science, Economics and Governance as Forms of Propagating Organization

> "I take informatics to mean the technologies of information as well as the biological, social, linguistic and cultural changes that initiate, accompany, and complicate their development (Hayles 1999a, 29)".

Katherine Hayles' quote indicates that there is a link between biological, cultural and linguistic information. It was also noted earlier and in POE that language and culture like living organisms also propagate their organization and hence their information. This also includes science, technology, economics and governance which are part of culture and will be treated separately because they provide vivid examples of propagating organization. The information that langauge and culture represent like biotic information is not Shannon or selective information but rather information with meaning, namely MacKay structural information.

Cultural and linguistic information is not fixed but depends on the context—as conditions change so do languages and cultures. This statement applies to the various sub-division of culture that we have explicitly identified, namely, science, technology, economics and governance. These forms of information do not represent Shannon selective information but rather MacKay structural information because of their dependence on context. Each one is more than a string of alphanumeric symbols or a string of 0s and 1s.

Let me provide an example of how linguistic meaning depends on context based on my experience of being the father of four children who in turn have provided me so far with four grandchildren. The meaning of the term Dad has changed for me over my lifetime. Dad used to be my father and then when I had children it meant me and then when my children had children and I became grandpa and Dad became the father of my grandchildren.

The point is that the meanings of words are context dependent. This is why I (Logan 2006 & 2007) identified words as strange attractors. They are strange attractors because the meaning of a word is never exactly the same as its meaning changes ever so slightly each time it is used because the context in which it is used is never the same. To illustrate the idea let us consider the word water which represents the water we drink, wash with, cook with, swim in, and that falls as rain, melts from snow, constitutes rivers, lakes, ponds and oceans, etc., etc. The meaning of water in each of these contexts is slightly different but there is a common thread and hence the claim that the word "water" acts as a strange attractor for a diverse set of contexts involving water.

A language is an organization of a set of symbols whose semantics and syntax is a form of information. A similar claim can be made for a culture, which Geertz (1973, 8) also defines in symbolic terms.

Information as a form of organization for either language or culture, although it is symbolic like Shannon information, still cannot be associated with Shannon information because linguistic and cultural information is context dependent and meaningful. It is also the case that language and culture are like living organisms in that they evolve in ways that cannot be predicted. We may therefore use the same core argument we did in POE to rule out the description of language and culture and their evolution with Shannon information. "The ensemble of possibilities and their entropy [for language and/or culture] cannot be calculated (Kauffman et al. 2007)." Therefore a definition of information as reducing uncertainty does not make sense since no matter how much one learns from the information in a linguistic or cultural system, as was the case with a biotic system, the uncertainty remains infinite because the number of possibilities of what can evolve is infinitely non-denumerable. Because science, technology, economics and governance are part of culture and it is also true that their evolution cannot be predicted and the argument we just made for language and culture applies to these subsets of culture as well.

At this point it is perhaps useful to define two forms of information micro-information consisting of isolated bits of information, the kind that are transmitted as Shannon information and are also components of a larger information systems or organization and macro-information or the organization of a system like a living organism, a language, or a culture. Other forms of macro-information include the specific elements of a culture such as a business, an economic system, a polity, science and the technosphere. Narrative is the organization of a text or a uttereance and therefore may be regarded also as a form of macro-information. Micro information is the string of characters and symbols that make up the narrative of a book, an article or a story.

There is still another property that the organzational information of language and culture share with living organisms that distinguishes them from Shannon information. This is the fact that language and culture, like life, are self-organizing phenonena and hence as is the case for biotic information and not the case for Shannon information we have a primitive model for the emeregnce of this information. Although we do not have a precise theory for how language and culture and the information and organization associated with them emerged we do have a number of proposals and models for how this might have happened through self-organization. Logan (2007) contains a review of these models.

The notion of organization as a form of information is based on the notion that the systems we have reviewed consist of components that are organized by some organizing principle. For living systems the components are the biomolecules of which living organisms are composed and the constraints or instructional information that allows the conversion of free energy into work is the organizing principle of these biomolecules, which is propagated as the organism replicates.

This model holds for languages where grammar is the organizing principle and the components are the individual words or semantics. Replication takes place as children learn the language of their parents or care givers.

The model also holds for social systems where the culture as patterns for behavior is the organizing principle and the components are the behaviors and judgments of the individual's of the society. Replication occurs as young people learn the intricacies of their culture from a variety of sources including parents, teachers and peers.

For technology the technosphere is the organization and the components are the individual inventions or artifacts. Replication takes place each time an inventor or innovator makes use of components of the technosphere to create a new artifact or invention.

The model holds for economic-governance systems where the economic model is the organization and the components are the individual business transactions. Examples of different economic models based on the work of Johnson and Earle (1987) are:

- individual families as basic economic unit;
- the big man tribal economic unit where the big man is the co-ordinator of economic activity and serves at the pleasure of the people;
- the chief dominated tribal economic unit where the chief controls all the means of economic activity but answers to a tribal council;
- the state or manor economy where the monarch or the lord of the

manor is the absolute ruler; as was case with medieval manor system, Czarist Russia and France before the revolution;
- the market driven system, which is democratic as in a republic like the USA or constitutional monarchy like the UK;
- the socialist state where private enterprise is controlled; and
- the communist state, which is state capitalism as was case with Soviet Union and Maoist China. China is now evolving into a mixed communist-socialist state.

The replication of economic-governance systems is through cultural and legal systems.

The model holds for science where the scientific method is the organizing principle and the components are the individual scientific theories. Replication occurs through the publication of scientific results and the education of new scientists.

Conclusions

We have demonstrated the relativity of information by showing that information is not a unitary concept independent of the phenomena it is describing or the frame of reference with respect to which it is defined. In particular we have shown that Shannon information cannot properly describe living organisms, language, culture and the various components of culture such as technology, science, economics and governance. We have examined the relationship of information to materiality, meaning and organization and showed that Shannon information is independent of meaning, organization and its material instantiation, which is just the opposite for biotic information, and the information associated with language and culture. We have also shown that that there exists an intimate relationship between information and organization for biotic systems and the elements of human culture including language, technology, science, economics and governance.

東館 5F
06.6241.0660
SPORTS ENTERTAINMENT!
ROUND1
ROUND1
心斎橋店
金・土・祝前日24時間営業
心斎橋筋
SHINSAIBASHI-SUJI
Step
心斎橋店
KoKuMIN

Chapter Four

Culture and Information and their Interrelation

Culture consists of the symbolic information that acts as an adaptive mental tool and is unique to humans. Culture is the mechanism whereby the learning of previous generations are passed on to the next generation through communication and social interactions.

As suggested by Boyd and Richerson (1985, 14):

> Individual learning ... can be costly and prone to errors. Learning trials occupy time and energy that could be allocated to other components of fitness, and may entail a considerable risk to the individual as well. Because of these costs, the investments of individuals in determining the locally favored behavior must be limited, and individual learning can lead to errors. Individuals may fail to discover an adaptive behavior, or a maladaptive one maybe retained because it was reinforced by chance. When these costs are important, selection ought to favor shortcuts to learning—ways that an organism can achieve phenotypic flexibility without paying the full cost of learning. Cultural inheritance is adaptive because it is such a shortcut. If the locally adaptive behavior is more common than other behaviors, imitation provides an inexpensive way to acquire it.

Geertz's (1973, 8) definition of culture emphasizes the symbolic nature of cultural information. He defines culture as "an historically transmitted pattern of meanings embodied in symbols, a system of inherited conceptions expressed in symbolic forms by means of which men communicate, perpetuate and develop their knowledge about and attitudes towards life." He goes on to add, that "culture is patterns for behavior not patterns of behavior."

Adam and Eve, Gossaert Thyssen

Culture is an extrasomatic form of instruction that provides individuals with an additional margin of survival. Culture is extra-genetic and plays a role similar to genetically transmitted instincts. Both genetically controlled instinctual behavior and culturally constrained behavior evolve with changing conditions. Instinctual behavior and culture both support survival. Without a culture a human being or a family unit for that matter would be unable to survive. If the environment undergoes a dramatic change the instincts that were inherited from a previous time could be detrimental to survival and they will certainly undergo a change and evolution if the species is to survive. The same is true of culture or else the society will not survive. There are in fact many historical examples of inflexible cultures that were unable to adapt to changing conditions and as a result did not survive. The culture of hunting had positive survival benefits and made for an easy life until game was depleted by over hunting. When this happened the hunters/gatherers supplemented their wild food with domesticated plants and/or animals. Hunting cultures evolved either into pastoral societies in which animals were not slaughtered to extinction but domesticated and culled in a controlled way or into agricultural societies in which plants were carefully cultivated and harvested. These activities required much more effort than hunting and as documented in Genesis humankind was driven out of the Garden of Eden and had to earn their bread by the sweat of their brows.

The Relationship of Language and Culture

Durham (1991, 8) claims that culture consists of "symbolically encoded" concepts which means culture is very much like language, which also consists of "symbolically encoded" concepts, namely, words. As a result many of the characteristics that we have discerned and posited for language may well apply to human culture. Language is both an explicit part of culture and the medium for its transmission.

Culture has an enormous impact on human thinking and therefore the mind is not merely an extension of the brain due to language but we need to add culture to the equation. Our new formulation for the mind is therefore:

mind = brain + language + culture.

I formulated my notion of the extended mind before reading Clark's (1997, 2003, 2008) formulation of the extended mind in which he claims both language and culture provide cognitive scaffoldings that extend the mind. Our ideas are parallel as he recently acknowledged (Clark 2008) and I wish to do likewise. Just as language provides a framework for conceptualization culture does the same thing as it stores all of the lessons that a society has acquired over the years. Given that language is a cultural artifact it makes sense that other cultural artifacts and processes would also contribute to the way the human mind is constructed.

Is Culture an Organism?

In the remainder of this chapter we will examine the possibility that culture, like language, evolved as an organism that was easy for the human mind to grasp and as a result gave rise to Universal Culture just as language evolved in such a way as to give rise to Universal Grammar.

Culture is essentially symbolic—a set of ideas, beliefs, information and knowledge. If it is to be transmitted and hence survive it must be easily acquired by the human mind as is the case with language. It is therefore logical to posit that culture like language evolved in such a way as to be easily acquired by humans. I have therefore suggested that Christiansen's (1994) idea that language is an organism can be extended to culture which may also be regarded as an organism, an obligate symbiont. If we accept this hypothesis then it follows by analogy that many of the conclusions Christiansen reached regarding language would apply to culture as well.

If we transform a paragraph of Christiansen, Dale, Ellefson and Conway (2001) that I quoted in the last chapter by replacing the word "language" with the word "culture", we arrive at some interesting thoughts about the nature of culture and its evolution. With this substitution Christiansen's (1994) notion of "language as an organism" can be extended to culture, which can also be considered as an organism in the same metaphorical sense.

> *Culture* exists only because humans can learn, produce, and process them. Without humans there would be no *culture*. It therefore makes sense to construe *cultures* as organisms that have had to adapt themselves through natural selection to fit a particular ecological niche: the human brain. In order for *cultures* to "survive", they must adapt to the properties of the human learning and processing mechanisms. This is not to say that having a *culture* does not confer selective advantages onto humans. It seems clear that humans with superior *cultural* abilities are likely to have a selective advantage over other humans.... What is often not appreciated is that the selection forces working on *culture*

> to fit humans are significantly stronger than the selection pressures on humans to be able to use *culture*. In the case of the former, a *culture* can only survive if it is learnable and processable by humans. On the other hand, adaptation toward *culture* use is merely one out of many selective pressures working on humans (such as, for example, being able to avoid predators and find food). Whereas humans can survive without *culture*, the opposite is not the case. Thus, *culture* is more likely to have adapted itself to its human hosts than the other way around. *Cultures* that are hard for humans to learn simply die out, or more likely, do not come into existence at all.

Christiansen, Dale, Ellefson and Conway (2001, 144–45) quote has been altered by substituting the word *culture(s)* for *language(s)*. We therefore conclude that culture like language can also be regarded as an organism that evolved to be easily acquired and preserved.

Culture Organisms belong to Individuals and not to a Society as a Whole

Each individual in a society is a symbiont with its language organism and its culture organism. The culture of the society is the species of all the individual cultural organisms in the society.

> People learn as individuals. Therefore, if culture is learned, its ultimate locus must be in individuals rather than in groups.... If we accept this, then cultural theory must explain in what sense we can speak of culture as being shared or as the property of groups ... and what the processes are by which such sharing arises (Goodenough, 1981, 54).

This insight of Goodenough justifies our assignment of the living organism to the culture of each individual and the culture of the group as a species of the conspecifics of individuals' cultures, The culture of each individual in a society can be quite different because there are components which depend on the family they are members of, the locale and country in which they live, their profession, the company or organization for which they work, their religious beliefs, their hobbies and a large number of other factors.

Kauffman defines a living organism as "a system of chemicals that has the capacity to catalyze its own reproduction (Kauffman 1995, 49)." Generalizing Kauffman's definition and applying it to culture we define culture as a system of symbols, ideas, beliefs and knowledge that has the capacity to catalyze its own reproduction. If we consider the culture of each individual as an organism then we may regard culture reproducing itself each time a child acquires a culture similar to his or her parents and other cultural conspecifics. But the child modifies their parent's culture as a result of the different influences that effect them coming from their peers. A process of selection takes place as those cultural elements which best serve the individual and their society predominate. The cultural organism of each individual in the society thus evolves along the lines of the Darwinian formula of descent, modification and selection. Then as was the case with the language of individuals the inheritance or descent is not by diploidy but that of polyploidy. The culture possessed by each individual can be characterized the way Christiansen and Ellefson (2002) characterized language, namely as "a kind of beneficial parasite—a symbiont—that confers some selective advantage onto its human hosts without whom it cannot survive."

The culture of the society as a whole is not an organism because it can not reproduce itself, rather it is a cultural species made up of the cultural organisms of all the individuals comprising the society. Just as conspecifics of a biological species are able to reproduce among themselves the conspecifics of a cultural species are able to communicate with each other, to cooperate, to collaborate, and share certain basic values and assumptions. One can speak of English and French cultural species and American, English-speaking Canadian and British cultures as subspecies of English-speaking culture. There are even finer gradations of subspecies within these three countries depending on vocations, hobbies, religion etc. They are subspecies in that they are distinct in some ways but they share certain common values of their cultural species. This is similar to the case of biological subspecies are that distinct but whose members can interbreed.

The culture that belongs to the community or society rather than the individual evolves through the mutations that arise in the idiosyncratic use of and modification of culture by individuals. Those idiosyncratic mutations can then be transmitted through the society by being incorporated into the individual cultures of other individual members of the cultural community.

We have developed two meanings to culture. The culture of the individual, an organism, and the common culture of a society, a species. The cultural community can be a nation state, a local region such as a city, a tiny village or a neighborhood, a profession, a community of practice, or even an extended family.

Universal Culture

Let me introduce another interesting and highly speculative notion based on regarding culture as an organism. Let us generalize Christiansen's (1995, 9) argument that in order to survive language evolved in such a way as to adapt itself "to fit the human learning and processing mechanism." This mechanism led to the universality of the characteristics of human language or to

Chomsky's Universal Grammar (UG). If natural selection acting on language as an organism led to the UG then we should expect natural selection acting on culture as an organism should lead to a universal set of rules that govern the social interactions within a culture which we will identify as Universal Culture (UC), defined as the set of universal elements which characterize all human cultures. The universals include such elements as: language, kinship relations, marriage, gossip and incest taboos.

Universal Culture and Universal Grammar have certain parallels as pointed out by Robin Fox (1989, 113), who makes a distinction between the process that represents an universal and the content of the universal or the way it represents itself,

> They (referring to cultures) may be unique at the level of specific content—like languages—but at the level of the *processes* there are remarkable uniformities—like language again.... Each outcome of a universal process can look very different. But it is nowhere written that universal processes should have identical outcomes.

The notion of a universality of human culture, however, runs counter to the main stream of the field of anthropology where the traditional focus has been on the description of primitive and exotic cultures and uncovering the variety and diversity of human culture. There are those that disagree and argue that there are more things in common than the things that are different. They claim that the basic structures of human culture are actually very similar and it is only the details that are different such as Fox who we have just cited.

Lee Cronk for one suggests that world cultures may be like the world's languages where there are many differences but an underlying common structure exists. He cites as evidence for this position Donald E. Brown' (1991) book *Human Universals* and the chapter entitled "Universal People" which details

> universals appearing in everything from the details of language and grammar to social arrangement to the ubiquity of music, dance, and play. The list includes some surprises. Every society has gossip, all societies understand the idea of a lie, they all have special types of speech for special occasions, they all use narrative, and they all have poetry with lines that take about three seconds to say. Men are everywhere on average more aggressive and likely to kill than women, though individual men and women do differ significantly from the average. Everyone has taboos on certain statements and certain foods. All societies are at least aware of dancing (though it is prohibited in some of them) and have some sort of music. Remarkably, everyone has children's music. If as cultural determinist dogma would have it, culture is all-diverse and

> all-powerful, why are there any such universals? Why aren't human cultures more diverse than they apparently are? Cronk (1999, 25)

Tiger and Fox (1971, cited by Brown 1991, 81) "argued that the important universals are not at the 'substantive' level where anthropologists usually seek them, but at the level of 'process' .... Processes may be universal even though their results are highly variable."

A Catalogue of Cultural Universals

Brown (1991, 130–41) has catalogued all those aspects of human culture which are universal or in his words are "near-universal." He asks, "what do all people. all societies, all cultures, and all languages have in common? (ibid., 130)" He attempts to provide an answer in terms of what he calls "the Universal People (UP)."

> The UP are aware of this uniqueness (i.e. their possession of culture) and posit a difference between their way—culture—and the way of nature. A very significant portion of UP culture is embodied in their language, a system of communication without which their culture would necessarily be very much simpler. With language the UP think about and discuss both their internal states and the world external to each individual.... With language, the UP organize, respond to, and manipulate the behavior of their fellows ... UP language is of strategic importance to those who wish to study the UP. This is so because their language is, if not precisely a mirror of, then at least a window into, their culture and into their minds and actions (ibid., 130).

Brown (1991, 130–41, 157–201) lists over one hundred items that human cultures right across the planet share in common on a universal or near-universal basis. Brown's list includes a number of universal features of culture associated with language including: prestige for good use of language; gossip; lies; humor; insults; and language change. "There are features of language at all basic levels—phonemic, grammatical, and semantic—that are found in all languages (ibid., 131)."

In addition to these features Brown lists the following set of universal or near-universal aspects of language: nouns and verbs; the possessive form; marking (good is never solely expressed as not bad); special speech for special occasions; narrative; poetry with a pause approximately every three seconds; figurative speech; metaphor; metonymy; onomatopoeia; gender; temporal duration; "units of time—such as days, months, seasons, and years"; cyclicity or rhythmicity; tense (past, present and future); similar classification categories ("parts of body, inner states, behavioral properties, flora, fauna, weather conditions, tools, space, and many other definite topics"); proper nouns; pronouns, first, second and third person; topographic and place names; antonyms and synonyms; numerals; kin terms distinguishing gender and generation; "semantic categories including motion, speed, location, dimension, and other physical properties"; words that are used more often are shorter; "binary discrimination" such as "black and white, nature and culture, male and female, good and bad, and ordered continua with a concept of a middle; measures and distances but not always with uniform units; taxonomies; logic terms such as *not, and, same, equivalent,* and *opposite*; symbols; conjectural reasoning; causality; subject/object distinction; mimetic elements such as hand signals; and gestures that can be mimicked, masked or modified and which are universally recognized".

In addition to these universal associated with language Brown also finds the following psychological and behavioral features of human culture universal: trial and error learning; a theory of mind; concept of self and others, self-awareness; understanding intentions; fear especially of loud noises, strangers and snakes; sexual attraction; homosexuality; flirting; jealousy; envy; recognition of others; prolific tool making and use (levers, containers, materials for tying, spears, weapons; and the use of fire); cooking; drugs; shelter; preparation for birth; post-partum natal care; group living such as the family; groundedness in a locality; marriage and courtship; adultery; family; child rearing; juvenile delinquency; traditional restraints on the rebelliousness of young men; nepotism; sex taboos; Oedipus complex; ascribed and achieved social status; social states; domination; prestige; labor division; male dominance, male rulers; male activities that exclude females; cooperative of labor; trade, gifts, food sharing; predicting and planning for the future; triangular relationships; government or public affairs; authority; power; collective decision making; leaders, never completely democratic nor totally autocratic; admiration of generosity; altruism; loyalty; rules; dispute settlement; proscription against rape, violence, and murder with sanctions; suicide; conflict; control of disruptive behavior; ingroup/outgroup classification; ethnocentrism; recognizing and employing promises; morality; values, ideals and standards; empathy; pride; shame; sorrow; need; daily routines; etiquette; hospitality; sex and excretion modesty; religion or belief in supernatural things; anthropomorphization; medicine; magic; divination; theories of fortune and misfortune; ritual; rites of passage; mourning; world view; dreams and interpretation; possessive case; property; rules of inheritance; aesthetic standards; art; imagination; story telling, narratives and myths; a need to explain the world; adornment; grooming; hair styles; dance; music (instrumental, vocal and children's); play; and games of skill and chance.

Some aspects of culture are near-universals including the domestication of dogs, notation systems, the association of poetry and ritual; the belief in spiritual entities such as the soul; the symbolism of red, white and black, capital punishment and abortion.

This list of universals comes from the literature and for the most part from the work of Brown (1991) who originally compiled the above lists. There is one universal that I believe should be added to the list which is a justice system to detect and punish cheaters. Although capital punishment is a near-universal almost every society has other forms of punishment for those that transgress against their society by cheating in one form or another.

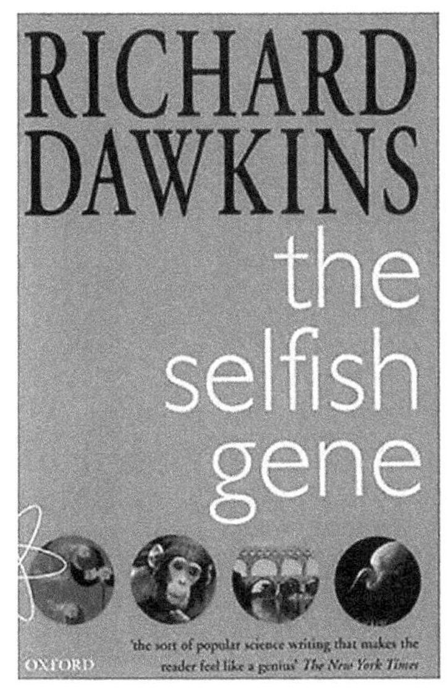

The Selfish Gene, Dawkins, 1976

Oxford University Press

Memes as the Replicators of the Organism of Culture

If culture is a symbolic organism, as we have posited, then its replication requires something analogous to genes, the replicators of biological systems. Richard Dawkins (1989) in his book *The Selfish Gene* has identified an analog to genes with his introduction of the meme as a cultural replicator. Dawkins considered the cultural meme as a way of extending Darwin's theory of evolution from biological systems to cultural or social systems.

> I developed the idea of the 'cultural meme' as a way of dramatizing that fact that genes aren't everything in the world of Darwinism.... The meme, the unit of cultural inheritance, ties into the idea of the replicator as the fundamental unit of Darwinism. The replicator can be anything that replicates itself and exerts some power over the world to increase or decrease its probability of being replicated (Dawkins 1996. 80–81).

Dawkins' notion of the meme, as a cultural replicator and the analog of biological genes helps us solidify our notion of culture as an organism. The meme not only accounts for the reproduction of culture but also natural selection as memes competes with each other memes for a place in human minds. Just as a biological organism can be defined in terms of its genetic composition so can a cultural organism be defined in terms of its memetic composition. The patterns of behavior that make up culture are all memes as are the words of a language and its grammatical structures.

What does it mean to propagate organization in the symbolosphere? Do memes, languages and cultures have purpose and intention? Institutions such as religions, social societies, nation states, a school of scholarly or scientific thought, have the collective purpose of its members and hence a purpose.

What is the mechanism that permits the replication of these memes? Well for one thing they exploit their hosts by providing a benefit, which assures their transmission or replication. Language and culture, however, are absolutely essential for human survival because of the dependency that has developed; just as modern society could no longer survive without electricity. There is a coupling of language and culture, for example, to the energy exploiting behavior of humans, which results in language and culture propagating their organization riding on the back of human metabolism.

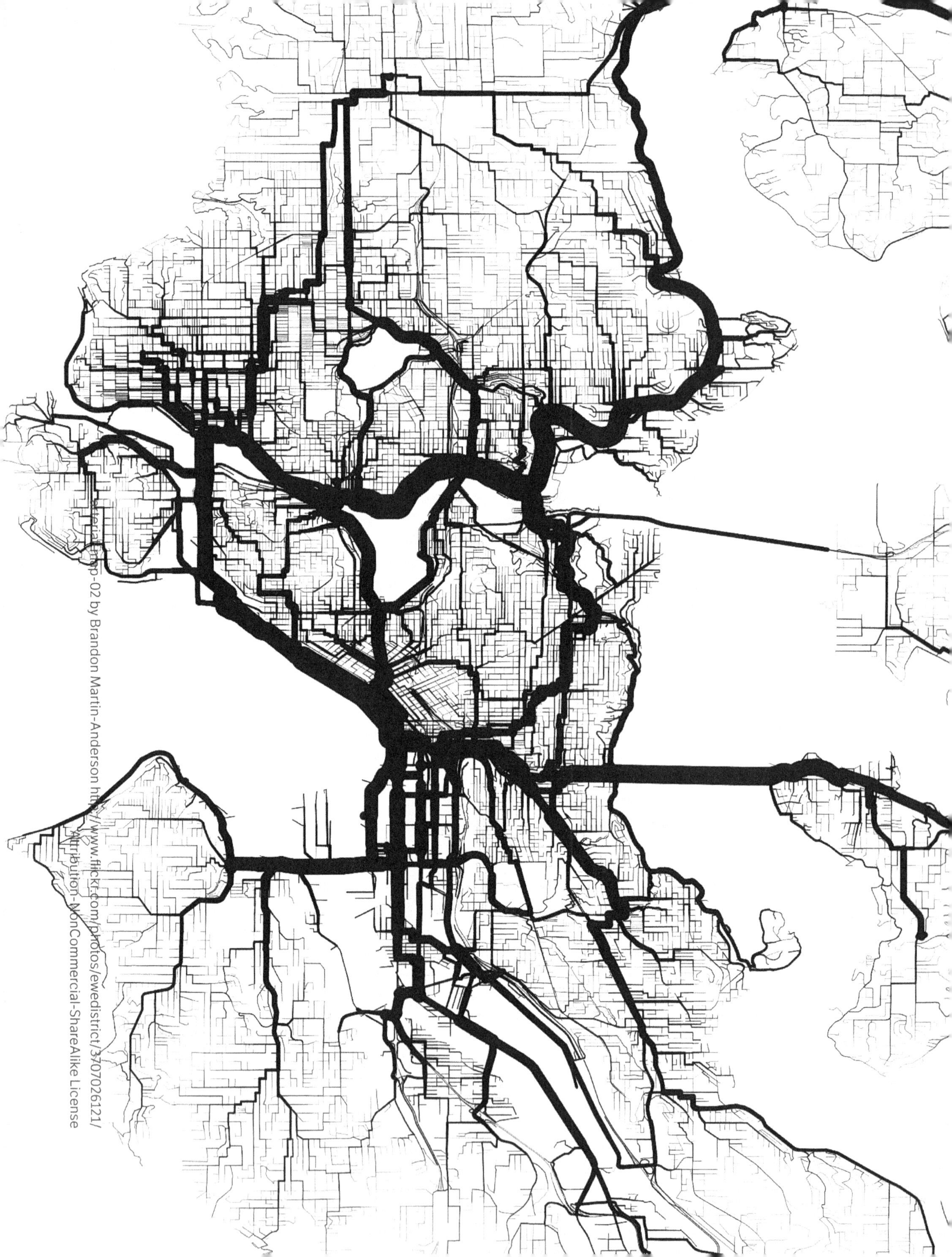
arterial map-02 by Brandon Martin-Anderson http://www.flickr.com/photos/ewedistrict/3707026121/
Attribution-NonCommercial-ShareAlike License

Chapter Five

Propagating Extra-Somatic Organization in the Symbolsphere

In Chapter 2 we reviewed the paper of Kauffman et al. (2007) hereafter referred to as POE in which it was shown that living organisms that occupy the biosphere propagate their organization materially by the constraints that allow them to channel free energy into work for their metabolism and replication.

These constraints are connected to information: in fact, simply put, the constraints are the biotic or instructional information that we have already defined.

The non-material, extra-somatic symbolic domain of human language and culture, which occupies a special place in the biosphere, was only cursorily identified in POE and not analyzed at all. The objective of this chapter is to consider these symbolic and conceptual aspects of human behavior, which comprise the symbolosphere, which is also described in this chapter in Section 2. We will analyze the way in which the elements of the symbolosphere propagate their organization. We begin with language and culture in Section 3 and then we treat three aspects of culture separately, namely, technology in Section 4), science in Section 5, and government and economics in Section 6.

Earth, Apollo 17, 1972

NASA

The propagation of organization in the symbolosphere is akin to the propagation of material organization in the biosphere as identified in POE. We treat the three elements of culture: technology, science, and economic-political systems separately because they provide vivid examples of propagating organization.

In his book *Investigations* Kauffman (2000) developed a number of interesting theses related to the propagation of organization by living organisms. In the final section of this chapter, Section 8, we extend to language and the different aspects of culture three of the properties Kauffman identified in *Investigations* for living organisms, namely the notions of

i. the exploration of the Adjacent Possible,
ii. the maximization of variety and hence Kauffman's putative fourth law of thermodynamics, and
iii. self-constructing systems.

Actually Kauffman included technology and economics in his analysis from time to time to illustrate these three notions. We will, however, attempt a more systematic approach to these three notions by including language, culture and science, which Kauffman did not deal with explicitly.

The Symbolosphere

In POE based on Kauffman and Clayton (2006) we argued that biology cannot be reduced to physics and that this implies that "the future evolution of the biosphere cannot be finitely prestated." In the same way that biology cannot be reduced to physics it is also the case that the symbolic conceptual aspects of human behavior, namely, language and culture cannot be reduced to, derived from or predicted from human biology. Nor can the future evolution of language and culture (the symbolosphere) be finitely prestated. The emergence of verbal language from mimetic communication described in Chapter 3 could never have been predicted from or reduced to the properties of mimetic communication. One could never have predicted the emergence of Proto-Indo-European nor its divergence into its many descendants such as English, Sanskrit, Greek, Latin, Italian and Romanian. Nor could one prestate the preadaptations of the cultures of the world, their technologies, economies and forms of governance, all of which depended on the physical environment they found themselves in among other factors. Nor can we prestate or predict the development or evolution of science and mathematics.

The symbolic domain of human language and culture is a product of human conceptual thought (Logan 1997, 2000, 2006a & 2007) and represents emergent phenomena and also, as we will show, propagating organization. They differ from the phenomenon in the biosphere that was the focus of the analysis in POE in that they are abstract, conceptual and symbolic and they are not materially instantiated nor do they have extension with the exception of technology. In the case of technology it is the symbolic concepts and organization that goes into the creation of the physical tools that propagates not the actual physical tools themselves. Another motivation for consideration of the propagation of organization represented by language and culture is the fact that the rate of linguistic and cultural evolution far outstrips the rate of human biological evolution and is therefore essential for understanding the evolution and development of the human experience.

The notion of the biosphere was introduced to distinguish it from the abiotic part of the physical universe or physiosphere. The biosphere as we have already indicated consists of living organisms, which represent a level of complexity above and beyond that of the abiotic part of the physiosphere and as such are emergent phenomena. For the purposes of our analysis we would like to suggest that the biosphere contains a more complex and emergent domain, the symbolosphere. The notion of the symbolosphere was first introduced by John Schumann (2003a & b) and later elaborated in Logan and Schumann (2005) and Logan (2006b). The symbolosphere is defined as the human mind and all the products of the human mind including symbolic abstract thought, language and culture. The universe constructs itself from energy, the biosphere constructs itself from biomolecules in the physiosphere and the symbolosphere constructs itself from concepts acting as strange attractors in the human brain for neural-based percepts as described in Section 2.

The part of the symbolosphere represented by the human mind is distinguished from the brain and is the domain of conceptual thought made possible by language. In Logan (2000 and 2007) mind is playfully defined using the formula: mind = brain + language. In this model the brain is part of the physiosphere and is basically a percept processor. It is only with language that conceptual thought by the human mind becomes possible as was described in Chapter 3.

By culture we will make use of Geertz's (1973, 8) notion that culture is embodied symbolically. All of the elements of culture are products of human conceptualization and represent emergent phenomena.

The symbolosphere is embedded in the biosphere and emerged from it just as the biosphere is embedded in the physiosphere and emerged from it. The symbolosphere includes aspects of human symbolic thought and culture such as, language, technology, science, governance, economy, writing, mathematics, computing, the Internet, poetry, music, and the visual arts all of which represent propagating organization. We will restrict ourselves in this analysis for the sake of brevity to the first elements of human culture in the above list, namely, language, technology, science, governance and economy.

With these definitions or this taxonomy we see that there have been three distinct levels of diversity and hence symmetry breaking in the history of the universe since the big bang, namely the emergence of a non-symmetric physical universe, the emergence of life or the biosphere on this planet and perhaps elsewhere in the universe and finally the emergence of the symbolosphere in the form of abstract generative symbolic language and culture among humans on earth and possibly elsewhere in the universe with other forms of life possessing symbolic intelligence.

Although the forms of organization in the symbolosphere are extra-somatic, non-material, and non-extensive they are, however, instantiated in some physical medium and represent human behavior and thought. Spoken language requires a number of organs of the human body for production and reception and the physical medium of air for transmission, which is sometimes enhanced through electronic devises such as the telephone or the microphone. Technology, science, governance, economy and all other aspects of culture are conceptual and symbolic. They are forms of organization that are physically instantiated in the material things they shape and/or control through downward causation.

All of the extra-somatic and non-material forms of organization that we will consider in this chapter originated basically with humans, although there is good evidence that some aspects of human culture emerged earlier in the Homo genus with Homo habilis, Homo erectus or Homo neanderthalis. A debate still rages as to whether earlier forms of genus Homo were capable of language. They certainly had a primitive form of culture, as they were toolmakers. We shall avoid this controversy, as it does not bear on the central theme of this chapter, namely the existence of the propagation of extra-somatic organization in the symbolosphere.

There is also a debate as to what extent animals that are not of the genus Homo also have culture and language. There is certainly evidence that some primates have a very crude culture and an equally crude use of technology as they propagate certain behaviors that enhance their survival such as using a long thin stick to extract termites from a termite nest or using rocks to crack open nuts. There is also the case of the monkeys on a Japanese island that wash their potatoes before eating them. The issue is whether or not this primitive or rudimentary form of culture is symbolic. I believe it is not. As far as language goes only humans seem to possess a fully generative language. The best that non-human feral animals can do is communicate with a small set of signals of not more than 50 in number.

It is also the case that human technology is far more sophisticated than any use of tools by non-human animals. Only humans use fire. Only humans use tools to make other tools. Only humans have organized their knowledge and developed science. Our focus will therefore be on the propagation of organization through human language and culture leaving the discussion of non-human instances of this, if any, to those more expert in the behavior of non-human animals.

The importance of the consideration of the symbolosphere is that we humans are, in the words of Terry Deacon (1997), "the symbolic species". What this means is that we are the only species capable of conceptualization and symbolization, i.e. of dealing with or processing information about an object or source that is not present to our senses in either space or time. Only humans are able to enter into a semiotic relation with an abstract symbol, i.e. a sign that symbolically stands for a concept or something that we cannot immediately apprehend or sense.

The Propagation of the Organization of Human Language

Based on Schumann's (2003a & b) work and the Extended Mind model (Logan 2000 & 2007) it has been postulated that the symbolosphere, human language and abstract symbolic conceptualization co-evolved and emerged at the same time (Logan and Schumann 2005). Human language is an emergent phenomenon and a complex adaptive system, which propagates its extra-somatic organization and evolves in a fashion very similar to that of living organisms as was described in Chapter 3. Another hypothesis that supports our hypothesis that human language propagates its organization is Dawkins' (1989) notion of the meme, which replicates elements of culture including language, and which we will treat in more detail in Section 4.

According to Kauffman (1995, 49): "A living organism is a system of chemicals that has the capacity to catalyze its own reproduction." Generalizing Kauffman's definition a language operates as a symbolic organism that has the capacity to catalyze its own reproduction. If we consider the language produced and comprehended by each individual speaker as a non-autonomous symbiont organism then we may regard language reproducing itself and propagating its organization each time a child acquires the language of his or her parents and other linguistic conspecifics.

By defining the language of each individual in the society as an organism not only do we meet Kauffman's criteria that an organism catalyzes its own reproduction but we are able to consider the evolution of this organism using Darwin's simple one line definition of evolution, namely, "descent with modification and selection." By descent Darwin meant reproduction, which is the acquisition of language by youngsters. One can now apply the concept of natural selection to the linguistic organism of each individual in a society, which undergoes modification by the way that individual uses the language making up new words or syntactical structures. Selection occurs when a neologism or a new use of an existing word catches on and is picked up by other speakers of the language.

Noam Chomsky

Catalytic closure, which Kauffman (1995, 50) has suggested is at the heart of the origin of life, might also provide a mechanism for the way in which language is reproduced from parents to children. If language exhibits the property of catalytic closure then the reproduction of some elements of the language catalyze the reproduction of others. Words are not isolated; they are part of a semantic web of meaning. The meaning of any given word is always given in terms of other words and therefore words catalyze each other and this is the sense in which a language exhibits catalytic closure.

By reproduction of the language we are talking about the "individual" language of each speaker and the process whereby young children are able to acquire the language of their parents with great ease. The biological capacity to imitate that hominids/humans acquired through biological selection plays the role of the analog of autocatalytic chemical reactions that create more of the same products. Catalytic closure is possibly the mechanism that allows acquisition of language to proceed so rapidly.

If we can accept the hypothesis that language is a non-autonomous symbiont organism that arises from catalytic closure we have a possible alternative to Chomsky's contention that the UG he formulated is hardwired. At the root of autocatalysis is self-organization or what Kauffman (1995) calls "order for free." "We have seen that the origin of collective autocatalysis, the origin of life itself, comes because of what I call 'order for free'—self-organization that arises naturally (ibid., 71)." If language emerged through a process of self-organization it comes with its UG already in place. The UG does not sit hard-wired in the brains of its users but rather it is an emergent property of the language itself, which replicates itself every time the symbiont language of the parent or caregiver reproduces itself as a symbiont language of the child. Kauffman's "order for free" translates into "grammar for free," the self-organization of the language itself. Put simply language as an organism evolved in such a way as to be easily acquired by an infant obviating the need to posit Chomsky's Language Acquisition Device.

Our definition of the reproduction of language as a living organism does not embrace Kauffman's (2000) definition of a living organism as an autonomous agent composed of biomolecules that is "able to act on its own behalf in an environment, (as) an autocatalytic system carrying out at least one thermodynamic work cycle". Language is clearly not a molecular system nor does it carry out a thermodynamic work cycle, but it does act on its own behalf propagating its organization. It does not have to perform thermodynamic work cycles, however, because it is a beneficial non-autonomous symbiont parasite that derives its energy from its host and in return increases the ability of its host to source and exploit free energy.

For the living organism that performs thermodynamic work it takes constraints to do that work and work to build those constraints. The constraints are built into the propagating organization of the autonomous agent by autocatalysis and have been identified in POE as instructional or biotic information to distinguish it from Shannon information.

For language the basic units are the words that comprise the semantics of a language whereas the constraints are the grammar or syntax. The autocatalysis of language arises from the fact that it takes concepts and grammar to make words and words to make concepts and grammar. Semantics and grammar are autocatalytic in the tradition known as the lexical hypothesis, which posits that "the lexicon is at the center of the language system (Donald 1991, 250, see also Levelt 1989 and Hudson 1984)." Because words can only be defined in terms of other words they form a semantic web as has been pointed out by Deacon (1997, 136).

> The metaphoric use of words and the way in which their various meanings interact can be likened to the web of symbol-symbol relationships that Deacon (ibid.) introduced to describe syntax. But the web of symbol-symbol relationships between different meanings of the same word create a semantic web of sorts which I suggest is the mechanism … to understand the evolution of words and the way language as an ecological system changes (Logan 2007).

In conclusion the analogy between living organisms and linguistic organisms consists of the following points:

- both propagate their organization;
- both evolve through descent, modification and selection;
- both are emergent phenomena;
- both arise from self-organization and catalytic closure; and
- both have a form of instructional information or constraints.

The analogy that both have a form of instructional information is less straightforward than the others and requires some explanation. For biotic agents instructional information provides the constraints necessary for guiding free energy into chemical channels so that work can be done to maintain and replicate the organism. The analogy for a linguistic organism is the grammar or syntax of the language, which constrains the flow of semantic elements to create meaningful propositions and to provide a structure of the language so that it can be easily learned by an infant and hence transmitted or replicated. We will call this form of instructional information linguistic instructional information to distinguish it from biotic instructional information and hereafter use the term instructional information as the generic term to refer to either form of instructional information.

The Propagation of the Organization of Human Culture

Culture is socially transmitted information, which takes the form of conceptual and symbolic mental representations in people's minds (Geertz 1973, 8). This means that culture is an extra-somatic and non-material form of organization that propagates from person to person.

Because culture propagates its organization and evolves like living organisms and language, we proposed in Chapter 4 that culture may be considered as a symbolic non-autonomous symbiont organism in the same manner in which we described language as a symbolic organism and symbiont ala Christiansen and Ellefson (2002).

Culture as an organism catalyzes its own reproduction. Each individual in the society, however, transforms or modifies the culture they inherit from their society to meet their own specific needs. Once again we have "descent with modification and selection" ala Darwinian evolution as the modifications of culture made by an individual are selected or ignored by the society based on their fitness. In this manner culture propagates its organization as described in POE.

We have argued that the culture embraced by individuals could be treated as a symbolic organism. As was the case with language culture is not an autonomous agent performing thermodynamic work cycles but rather a beneficial parasite, which pays for its consumption of energy by enhancing the ability of the individual and the society to which they belong to better source and exploit free energy. The relationship is symbiotic and is similar to that of those plants that play host to a fungus that fixes nitrogen and hence enhances the plants ability to transform sunlight into usable energy, which the plant then shares with the fungus.

For living organisms we identified the constraints operating on them as those of organic chemistry and chemical autocatalysis while for linguistic organisms we identified grammar or syntax as the form of instructional information operating in this system. With respect to culture the constraints are the social pressure to conform, which results in a more or less uniform behavior in a society. This uniformity does not apply in every case because of individual idiosyncrasies or rebellion. This provides the modification of the descent of culture from one individual in the society to another. It is by a process of selection that the cultural norms in a society change and evolve paralleling the evolution of living organisms.

The basic units of a culture are the patterns or models for behavior that comprise the individual's belief system. The constraints, on the other hand, are the social norms and social pressure of the society. The autocatalysis of culture is the fact that societies self-organize themselves.

In conclusion the analogy between living organisms and cultural organisms is similar to the one for linguistic and living organisms. They all propagate their organization; evolve through descent, modification and selection; are emergent phenomena; arise from self-organization and catalytic closure; and have a form of instructional information.

The culture of a society incorporates among other things its technology, science, economy and system of governance, which will be treated in Sections 5, 6 and 7 respectively. We now turn to an examination of these individually because they too represent propagating organization and display a pattern of evolution ("descent with modification and selection") very much like that of living organisms.

The Evolution of Technology

The emergence of technology almost certainly preceded language as evidenced by the fact that hominid toolmakers can be traced back to Homo habilis. The refinement of tools and their proliferation as well as the beginning of a technology-based culture, however, seems to have begun much later, about 50,000 years ago. According to Dunbar (1998, 105):

> Symbolic language ... would have emerged later as a form of software development ... probably at the time of the Upper Paleolithic Revolution some 50,000 years ago when we see the first unequivocal archaeological evidence for symbolism (including a dramatic improvement in the quality and form of tools, the possible use of ochre for decorative purposes, followed in short order by evidence of deliberate burials, art and non-functional jewelry). (ibid., 105)

The evolution of technology follows a pattern similar to that of living organisms as has been pointed out by a wide variety of authors. The first was the English critic and satirist Samuel Butler writing a mere four years after the publication of The Origin of the Species. More recent and more serious suggestions have been made by Basalla (1988), Mokyr (1990), Vincenti (1990) and Cziko (1995).

Basalla (1988) cites three basic analogies between technological and biological evolution. The first is the fact of the great variety of both biological organisms and technological tools. Basalla cites the fact that the U.S. Patent Office granted approximately 4.7 million patents between 1790 and 1988, the date of the publication of his book *The Evolution of Technology*. As he put it: "The variety of made things is every bit as astonishing as that of living things."

Basalla's second point is that technology evolves through a process of descent and modification: "Any new thing that appears in the made world is based on some object already in existence (ibid., 45)." He cites many examples of how innovative technologies borrowed significantly from earlier technologies citing the cotton gin, the electric motor and the transistor as three examples.

Gutenberg's moveable type printing press is another example. Gutenberg made use of the ideas of Laurens Janszoon Koster who had earlier built a block printing press in which a page was carved out of a block of wood in reverse. Koster also made use in some instances of movable type fonts also carved in wood. Koster's press was not original either but was borrowed from the block printing presses used in China, the idea for which was derived from the Chinese practice of printing patterns on silk cloth.

The third point that Basalla makes is that technologies survive through a selection process by which a society chooses a particular technology from a large number of variations for incorporation into its material life.

Mokyr's (1990, 275) approach to the evolution of technology is to consider the evolution of know-how rather than the physical artifacts:

> The approach I adopt here is that techniques..., namely, the knowledge of how to produce a good or service in a specific way—are analogues of species, and that changes in them have an evolutionary character. The idea or conceptualization of how to produce a commodity may be thought of as the genotype, whereas the actual technique utilized by the firm in producing the commodity may be thought of as the phenotype of the member of a species. The phenotype of every organism is determined in part by its genotype, but environment plays a role as well. Similarly, the idea constrains the forms a technique can take, but adaptability and adjustment to circumstances help determine its exact shape. Invention, the emergence of a new technique, is thus equivalent to speciation, the emergence of a new species.

Vincenti's (1990) approach to the evolution of technology was to develop a "variation-selection model for the growth of engineering knowledge." He suggests that the most efficient way to design new technology is to create variations vicariously and cheaply through modeling (either physical models or computer simulations) and then employ a selection process to pick the form of technology that will be finally built. Vincenti's focus like that of Mokyr is on know-how and also the most efficient way of achieving it through vicarious variation and selection.

Cziko (1995), who cites the work of Basalla (1988), Mokyr (1990) and Vincenti (1990), has created a Universal Selection Theory that includes the notion that technologies evolve in a manner similar to living organisms. "The adapted nature of technology and its progress is hard ... to doubt."

Finally I cite my own work in which I too saw the evolution of technology as analogous to that of living organisms:

> Cognitive tools and physical technology are two resources at the disposal of human innovators, and the needs or demands of society are often the motivating force. Necessity is the mother of invention, yet invention does not occur in a vacuum. All of the previous innovations in a culture provide the resources, both cognitive and physical, for the next level of innovation. The previous innovations also contribute to changes within the socioeconomic system that give rise to new social demands. Each new invention, technological innovation, or discovery gives rise to new technical capabilities, new cognitive abilities, and new social conditions. These then interact with the existing economic, political, social, cultural, technical, and cognitive realities of the culture to set the stage for the next round of innovation. Thus, technological change in our model is part of an ongoing iterative process. It began with the inception of Homo sapiens and continues to this day at an ever-quickening pace (Logan 2004b, 125).

The Evolution of Science

Science is another symbol-based activity unique to humans, which also propagates its organization. The mechanism for the propagation of science's organization is what Thomas Kuhn (1972) termed normal science. Every success in science gives rise to a paradigm, which is articulated and applied to as many phenomena as possible. This is the mechanism of descent. Once a paradigm fails to provide a satisfactory description of nature a period of revolutionary science begins with the search for a new paradigm. This is the mechanism of modification. If the new paradigm provides a satisfactory explanation to the science community by providing replicable results a new round of normal science begins. This is the mechanism of selection. Science propagates its organization through normal science and evolves by descent, modification and selection just like living organisms. The analogy between the Darwinian evolution of living organisms and the process of descent, modification and selection in Kuhn's model led him to cautiously conclude at the end of his analysis of scientific revolutions the following:

> The analogy that relates the evolution of organisms to the evolution of scientific ideas can easily be pushed too far. But with respect to the issues of this closing section it is very nearly perfect.... Succes-

> sive stages in that developmental process are marked by an increase in articulation and specialization. And the entire process may have occurred, as we now suppose biological evolution did, without benefit of a set goal, a permanent fixed scientific truth, of which each stage in the development of scientific knowledge is a better exemplar (Kuhn 1972, 172–73).

Karl Popper (1979, 261) whose description of science differs from that of Kuhn's, nevertheless also found an analogy between the evolution of science and that of living organisms:

> The growth of our knowledge is the result of a process closely resembling what Darwin called 'natural selection'; that is, the natural selection of hypotheses: our knowledge consists, at every moment, of those hypotheses which have shown their (comparative) fitness by surviving so far in their struggle for existence; a competitive struggle which eliminates those hypotheses which are unfit.

The Evolution of Governance and Economics

Because governance and economics are so intertwined and because economics by and large determines governance we will treat them together. Economic and political institutions propagate their organization and evolve in much the same way as living organisms and symbolic ones like language and culture through "descent, modification and selection".

Biological factors dominated the evolution of pre-human hominid and human existence at first. With the emergence of technology, language, and culture, these factors also played a key role in human evolution. Human biology and culture co-evolved (Boyd and Richerson 1985). "Population and technology have a feedback relationship; population growth provides the push, technology change the pull. But ... it is fundamentally population growth that propels the evolution of the economy (Johnson & Earle 1987, 5)."

Johnson and Earle (1987) identified the following stages of socialization that emerged with each incremental increase in population density:

1. *family-level groups*, which divided into either the family camp or the family hamlet;
2. *local groups* of 5 to 10 times the number of families of the family-level group, which came together for the purpose of defense or food storage;
3. *regional polities* that arose out of local groups and at moderate populations formed into a *chiefdom* and at large population levels into a *state*.

The individual units of governance and economy that Johnson and Earle identify, the family, the hamlet, the tribe headed by a big man, the chiefdom and finally the state are all forms of organization that propagate from one generation to another. With an increase in population due to the success of the economy at a lower level of organization a higher more complex level of organization emerges just as more complex biotic organisms emerge from simpler ones.

> As we have seen, at each evolutionary stage existing organizational units are embedded within new, higher-order unifying structures. Hamlets are made up of families, local groups of hamlets, regional chiefdoms of local groups, and states of regional chiefdoms. The earlier levels continue to operate but with modified functions. Thus the local group of a stateless society, which had formerly been a unit of defense, is transformed into a unit of taxation and administration as it becomes incorporated into the state. (ibid., 322)

Complexity, Emergence and the Evolution of Economic-Polities

As human societies succeeded in their ability to procure through hunting and gathering natural sources of food (and hence free energy) their population density increased, which led in the long run to a depletion of their food supply. The population overload led to new challenges and chaos. From this chaos far from equilibrium a new level of order emerged ala Prigogine (1997) in the form of the domestication of plants and animals. This pattern of domestication occurred throughout the world in isolated communities approximately 10,000 years ago at the end of the last ice age. While it is true that at the local level one society might learn domestication from its neighbors it is also true that agriculture and pastoralism emerged independently on every continent and in almost every ecosystem in the world. The explanation of the emergence of domestication out of the complexity of population overload parallels the strong emergence model described by Clayton (2004) and Kauffman and Clayton (2006).

The domestication of plants and animals led to new challenges and new levels of complexity, which in turn gave rise to new levels of increasing order in the form of family-level groups (camps and villages), local groups ('big man' systems) and regional polities (chiefdoms and states). Each new political system emerged from the population overload of the previous political system. It was a result of propagating organization through social and cultural transmission that the features of the previous economic-political system were incorporated into the new political order as was pointed out by Johnson and Earle (1987).

The Adjacent Possible, the Maximization of Variety and the Self-Constructing Symbolosphere

In his book *Investigations* Kauffman (2000) deals with the many levels of complexity of the material world but one level that was only dealt with cursorily was the non-material symbolosphere of language and culture. In this section we will extend to language and culture, i.e. the symbolosphere, Kauffman's arguments made for the biosphere. We shall attempt to expand Kauffman's notion that the universe, including the biosphere is constantly probing the Adjacent Possible and ever increasing the diversity of the symbolic universe by showing that the symbolosphere is also constantly probing its Adjacent Possible and as a consequence also increasing the diversity of the universe. We will also attempt to extend to the symbolosphere Kauffman's putative fourth law of thermodynamics, which states that self-constructing open systems like the biosphere maximizes the rate of creating diversity. And finally we will attempt to show that the symbolosphere like the biosphere is a self-constructing system. Kauffman has also argued that human economies and technology are also constantly probing the Adjacent Possible. We will extend this notion to all aspects of the symbolosphere, which are also constantly probing the Adjacent Possible.

The Adjacent Possible

A central thesis of *Investigations* is the existence of the Adjacent Possible in the biosphere, which Kauffman (2000, 22) defines in the following manner:

> Autonomous agents forever push their way into novelty—molecular, morphological, behavioral, organizational. I will formalize this push into novelty as the mathematical concept of an "Adjacent Possible," persistently explored in a universe that can never, in the vastly many lifetimes of the universe, have made all the possible proteins sequences even once, bacterial species even once, or legal systems, even once. Our universe is vastly nonrepeating; or … nonergodic.

We claim that there exists an 'Adjacent Possible' for the symbolosphere. In fact, Kauffman (2000, 54) acknowledges this for certain elements of the symbolosphere. "Science, technology, and art tumble into the Adjacent Possible in roughly equal and yoked pace." "The universe is vastly non-equilibrium, vastly nonergodic at the level of complex organic molecules. A fortiori, the universe is vastly nonergodic at the level of species, languages, legal systems and Chevrolet trucks (ibid. 145)." We claim that all elements of the symbolosphere are nonergodic. In the Extended Mind model (Logan 2007) words are regarded as representing concepts as strange attractors for the percepts associated with those concepts. Words are strange attractors because they never return to the same place in the configuration space of meaning because their exact meaning depends on the context of their use or the semantic web that surrounds their use. Since they are strange attractors they are nonergodic.

Kauffman (2000, 143) claims that "the biosphere has been expanding, on average, into the Adjacent Possible for 4.8 billion years" and as a result "there are now a standing diversity of 100 million species" with an estimated 10 trillion different genes representing a diversity that "is likely to be hundreds of trillions or more" organic chemical species.

The symbolosphere, on the other hand, has only existed by most accounts 50 to 150 thousand years (some will claim a million or two years) but has generated an enormous amount of diversity. There are extant some 6,000 languages not counting various local dialects. There are also many languages, which have become extinct. Most extinct languages leave no fossils with some exceptions like Proto-Indo-European or Latin that have diverged into many other languages and in the case of Latin have left a written record.

How many words in each language? English has approximately one million. Assuming the others have on average only 100,000 then the sum total of extant words in all the languages of the world is over half a billion words. But this is not the extent of the variety in the symbolosphere. We must also take into account all of the propositions or sentences that have been constructed from these words since the beginning of language. Let us assume a population of 6 billion people (we are only counting those alive today) with an average lifetime of 50 years uttering a hundred sentences per day. This yields some 10,000 trillion (10^{17}) sentences since the symbolosphere came into existence. Each year the number of sentences will increase by 200 trillion at today's population level. And the reckoning only takes into account spoken language. There is also all the variety created in the written word, technology, economics, laws, and cultural artifacts such as clothing, jewelry, art objects, etc.

Maximizing Variety and Fourth Law of Thermodynamics

As we saw in the last section by probing the Adjacent Possible "autonomous agents forever push their way into novelty" with the result that there is a "persistent evolution of novelty in the biosphere (ibid., 22 & 5)." The same dynamic holds in the symbolosphere, which for example increases linguistic novelty or variety in a number of ways including the creation of new words (neologisms) and new grammatical elements or structures through grammaticalization and by bifurcating into myriad accents, dialects and new languages such as the way Latin diverged into French, Italian, Spanish, Portuguese, Catalan, and Romanian. The symbolosphere is also increasing its novelty through the diversification of culture a fact Kauffman (ibid., 229) acknowledges for the economy: "The economy, like the biosphere, is about persistent creativity in ways of making a living." It is worth noting that the persistent economic creativity Kauffman identifies is in part due to conceptualization and the use of language.

Kauffman (ibid., 3–4) formulates a putative Fourth Law of Thermodynamics based on the persistent emergence of novelty in the Adjacent Possible for both the biosphere and the econosphere. "Biospheres maximize the average secular construction of the diversity of autonomous agents.... On average, biospheres persistently increase the diversity of what can happen next." Our claim is that this putative fourth law, if it is correct, applies with equal validity to all elements of the symbolosphere as is evidenced by the persistent novelty of technology, science, the law, literature, music, and the visual arts.

Self-constructing Systems

A central theme in *Investigations* (Kauffman 2000) is the notion that the universe and the biosphere are self-constructing systems. "A coevolving biosphere accomplishes (the) coconstruction of propagating organization (ibid., 5)." We wish to posit that the symbolosphere is also a self-constructing system. It takes thoughts or concepts to create words and words to create thoughts or concepts. Just as autonomous agents emerge in the biosphere through autocatalysis a similar mechanism works in the symbolosphere.

> The emergence of language and conceptual thought is an example of an autocatalytic process. A set of words work together to create a structure of meaning and thought. Each word shades the meaning of the next thought and the next words. Words and thoughts are both catalysts and products of thoughts and words. Language and conceptual thought are emergent phenomena. They bootstrap themselves into existence.
>
> It is impossible for us to determine because of the remoteness of the events which came first, the language skills, the social/communicative skills or the cognitive skills but one can argue that language, social/communicative skills and cognitive skills form an autocatalytic set of skills which reinforce each other (or bootstrap each other into existence) and which conferred upon those hominids that possessed them a reproductive advantage. (Logan 2007, 45. 173)

The driving force of the self-construction of the biosphere is autocatalysis, which Kauffman (2000, 37) attributes to a phase transition. He argues that, "as molecular diversity of a reaction system increases, a critical threshold is reached at which collectively autocatalytic, self-reproducing chemical reaction networks emerge spontaneously (ibid., 16)."

Let's extend this argument to the symbolosphere. Perhaps with the increased lexical/conceptual diversity of a protolanguage system (Bickerton 1998) (presumably the first form of human language in which there was only a semantics and no syntax) a critical threshold is reached at which collectively autocatalytic, self-reproducing symbolic networks emerge spontaneously with a full-blown syntax and grammar.

Some evolutionists suggest that it is difficult to explain cooperation and altruistic behavior in terms of natural selection because selfish individuals would have a selection advantage over altruistic ones. Various solutions to this problem have included group selection, kin selection and reciprocal altruism. A debate still rages as to which of these mechanisms if any can explain altruism. In *Investigations* Kauffman (2000, 75) makes an interesting attempt to resolve this issue by focusing on cooperation instead of altruism with its implication of sacrifice. "The central factors underlying (the) buildup of organization are the same factors that apply in an economy—that merely human extension of biospheres. The central factors, in fact, center on 'advantages of trade'."

We already saw the 'advantage of trade' at work in the example of symbiosis between a fungus and a plant where the fungus fixes nitrogen and absorbs energy from the root of the plant. A similar 'advantage of trade' can be used to understand the emergence and use of language, which is a beneficial parasite, a symbiont. There is also a mutual advantage to individuals exchanging information and coordinating activities that helps all participants. Consider the following example, which illustrates the 'advantage of information trade'.

The information I_a that cost agent A the work W_a to obtain can be shared with agent B at very little extra cost to either agent A or B. Let W_x be the cost to A to share I_a and W'_x the cost to B to obtain I_a from A. Agent A shares his information in the hope that agent B will reciprocate at some later time by sharing information I_b that costs agent B W_b to obtain plus W_x to share. It will also cost A W'_x to obtain I_b from B. Let us assume for simplicity that $W_a = W_b = W$ and $W_x = W'_x$. Then we can calculate the economic advantage of the exchange of information I_a and I_b between agents A and B. Let us assume two scenarios where A and B both obtain I_a and I_b: once by cooperation and once independently without cooperation.

With cooperation: The cost to agents A and B is the same, namely, $W + 2\,W_x$ for a total cost to the two of them of $2W + 4\,W_x$.

For the independent non-cooperation scenario: the cost to each for info I_a and I_b is 2W for a total cost of 4W to the two of them.

Given that $W >> 2W_x$ we immediately see the advantage of the cooperative scenario. We can conclude from this that in this model there is a natural advantage to cooperating and hence we have an explanation of the kind of organization that leads to altruism and how it is connected to language. We see that altruism can arise through natural selection in the same way that the plant and fungus formed a symbiotic relationship and a mutual economy to the advantage of both.

In *Investigations* Kauffman (2000) draws an analogy between living organisms interacting cooperatively and human economics. Given that language is part of the infrastructure of human economics it follows that language coevolved with human cooperation.

Non-human economics is conducted by non-symbolic signs or iconic and indexical signs. Natural selection and co-evolution give rise to symbiotic relationships and cooperation among and between species. Human economics, on the other hand, is conducted by language and culture or symbolic signs. Symbiotic relationships are conceived of and communicated through the symbolic activities of human language and culture. The conceptualization that language makes possible gives rise to a great variety of human economic systems that have allowed humans to populate almost every corner of the globe and has given rise to the domestication of plants and animals; manufacturing and most recently the knowledge economy.

Conclusion

The propagation of organization is not only a characteristic of living organisms but also, as we have shown, a number of abstract, symbolic, extra-somatic, non-material, non-extensive mental activities of humans in the symbolosphere including language, culture, technology, science, governance and economy. This result extends the results obtained in POE in which the propagation of organization was demonstrated in the material abiotic and biotic worlds. It also indicates a universality of the propagation of organization and the emergence of more complex forms of organization from simpler ones.

To understand the true nature of the evolution of humans we need to consider the coevolution of two domains:

1. the physically instantiated human body including the brain
2. the non-extensive symbolosphere of the human mind and all of the products of the human mind including abstract symbolic thought, language, culture, the technosphere, science, governance and economics.

There is a symbiotic relationship between these two domains and a parallel development. Both domains constantly probe their respective Adjacent Possibles. Both domains maximize their variety as predicted by Kauffman's (2000) putative fourth law of thermodynamics. As a matter of fact the symbolosphere seems to increase its variety at a much faster rate than the human body. And finally both domains are self-constructed systems as suggested by Kauffman.

A Highly Speculative Postscript – A Possible Bridge between Shannon and Biotic Information?

In POE reviewed in Chapter 2 we showed that biotic instructional information that informs or instructs living organisms is quite distinct from Shannon's classical definition of information as negentropy. If language and culture propagate their organization in a manner similar to that of autonomous biotic agents, i.e. living organisms, perhaps there is some common feature(s) that are shared by biotic or instructional information informing biotic systems and Shannon information informing human symbolic thought.

Let us start with the notion that materially instantiated instructional or biotic information informs or instructs the biomolecular components of a living organism how to behave through biochemical processes. Shannon information operating in the context of linguistic and cultural systems informs or instructs the human psyche through non-material symbols. Is there a sense in which Shannon information can be considered a form of instructional information? Shannon information informs or instructs the receiver of what information is being sent by the sender. If we accept these definitions then we can define a generalized instructional information that embraces both instructional or biotic information defined for living organisms and Shannon information defined for human symbolic communication. This seems like a natural complementarity as the term information implies that someone or something is being informed and hence instructed. Therefore all forms of information are instructional but the precise nature of the instructional information is determined by context, i.e. by the nature of the recipient of the information, hence the distinction between Shannon and biotic information.

In POE we suggested that information was not an invariant independent of the frame of reference in which it operates but it depended on the context in which it is used. This statement is still correct but there is one common aspect of these two different forms of information we have identified which is that they both inform by definition and hence they both instruct. Biotic information instructs the cell how to convert free energy into work needed for growth and replication. The human symbolic information, i.e. Shannon information, contained in language and culture performs a similar function in that it affects human activity in such a way as to enhance the way in which sources of free energy can be found and converted into useful work. The purpose of language and culture is ultimately to enhance the ways in which human can source energy and perform work and ultimately enhance human propagation.

In closing this chapter I wish to acknowledge that the ideas presented here were stimulated by my POE co-authors Stuart Kauffman, Bob Este, Randy Goebel, David Hobill and Ilya Shmulevich. I must give special mention to Stuart Kauffman. whose wonderful books have always stimulated me and who spent time with me, often by telephone, discussing these ideas.

Chapter Six

Propagating Organization, Neo-Duality and Material and Non-Material Emergence

The neo-duality picture that we will develop in this chapter is richer that the physicalist view that all phenomena in the world can be explained in terms of basic physics.

When I first began my studies as a physics student I too thought all phenomena could be explained ultimately by physics. Fortunately I grew out of this point of view as I discovered the variety and complexity of the phenomena of my world.

Stuart Kauffman (2000) in his book *Investigations* introduced the notion of propagating organization as a new union of matter, energy, work, constraint and information exemplified by the vast organization of the coevolving biosphere. In POE reviewed in Chapter 2 Kauffman et al. (2007) studied propagating organization in the material abiotic and biotic worlds. In the last chapter Kauffman's notion of propagating organization was extended to the non-material symbolic domain as exemplified by human language, culture, science, technology, governance and economics. It was posited there, as it was in POE and Kauffman and Clayton (2006), that the transition to higher orders of organization can only be explained in terms of strong emergence as described by Clayton (2004) in *Mind and Emergence*.

Clayton describes three basic schools of thought with respect to the question of the relationship between higher orders of organization and the components out of which they are constructed and from which they emerge. The three schools according to Clayton consist of physicalists, dualists and emergentists. The emergentists represent a third option between the physicalists and the dualists according to Clayton. The physicalists believe that all phenomena and all things that exist are basically physical or material and that ultimately everything can be and will be explained in terms of basic physics. The dualists on the other hand believe that in addition to the physical world there is also another element, which is "a soul, self, or spirit that is essentially non-physical (ibid., v)." Clayton citing el-Hani and Pereira (2000, 133) describes the emergentist position as consisting of following four elements:

1. All things are made of the basic particles described by physics and their aggregates;
2. As aggregates gain a level of complexity novel properties emerge;
3. These properties cannot be reduced to or predicted from the lower level from which they emerged; and
4. Higher-level entities causally affect the lower level entities from which they are composed and from which they emerged in what is called downward causation.

Clayton also identifies two major divisions within the emergence school of thought namely the strong and weak emergentists. Clayton, a strong emergentist himself, describes strong emergence as the belief that the new higher levels of complexity that emerges are ontologically distinct from the lower levels from which they come and that physics will never be able explain these higher level phenomena. The weak emergence position is that, yes, the levels are distinct but that ultimately they can be reduced to physics once a deeper understanding of the world is achieved.

A Comparison of Material and Non-material Emergence

Human symbolic interactions are naturally part of the human biotic system and hence are part of the biosphere. We choose, however, to make a distinction between the purely biological interactions of biosemiosis, on the one hand, and human language and culture, on the other hand. Biosemiosis is the communication of information instantiated in the biomolecules and organs of which living organisms are composed where the information that is communicated is not symbolic, i.e. standing for something else. It is therefore the case that the information cannot be separated from those biomolecules or the transmitters or the organs in which they are instantiated. DNA does not symbolize RNA but contributes to its creation chemically through catalysis. The same is true of RNA, it is not a symbol of the proteins it helps to create—it actually catalyzes their chemical composition. The neuronal signals are not symbols of something else but are actual physical signals. The medium and the information content or messages of biosemiosis is the same. Human language and culture, on the other hand are symbolic in which the information is not instantiated materially but is only physically mediated and as a result are able to move from one medium to another.

We make a distinction between material and non-material emergence. Examples of material emergent phenomena include regular hexagonal convection cells, weather patterns in the abiotic world and living organisms in the biosphere. Non-material emergent phenomena include human language, conceptual thought and culture all of which belong to the symbolosphere. The symbolosphere, originally introduced by Schumann (2003a & b), consists of the human mind and all the products of the mind, namely, its abstract thoughts and symbolic communication processes such as spoken and written language and the other products of the human mind and culture such as music, art, mathematics, science, and technology.

Non-material emergence differs from material emergence in that the first of the four elements el-Hani and Pereira (2000, 133) used to describe emergence does not hold, namely that all things are made up of basic particles. Human language, conceptual thought and culture are not made up of basic particles described by physics, they have no extension and they exist in the symbolsphere and not the 6N (where N is the number of particles in the system) dimensional configuration space of physical particles.

As was argued in the last chapter and has been argued by Kauffman (2000) and Clayton (2004) biology cannot be predicted from or reduced to physics. In the same way that biology cannot be reduced to physics it is also the case that the symbolic conceptual non-material aspects of human behavior, namely, language and culture cannot be reduced to, derived from or predicted from the biology of the human brain and the nervous system from which they arise. The symbolic domain of human language and culture are a product of human conceptual thought (Logan 2000, 2006a & 2007) and represent emergent phenomena and propagating organization. They differ from living organisms that populate the biosphere in that they are abstract, conceptual and symbolic and not materially instantiated as such with the exception of technology. In the case of technology it is the concepts and organization that goes into the creation of the physical tools that are emergent and propagate not the actual physical tools.

Neo-dualism

It is because of the existence of non-material emergence and the symbolosphere that the notion of neo-dualism was introduced in Logan and Schumann (2005) and extended in Logan (2006b). While carefully distinguishing the different forms of emergence Clayton (2004) did not entertain the possibility of different kinds of duality. Neo-dualism is quite different than the dualism that Clayton (2004, v) defines, a dualism that incorporates the notion of soul or spirit. "Dualists believe that ... humans consist of both [a] physical component and a soul, self, or spirit that is essentially non-physical (ibid.)."

We agree with dualists that there is a non-physical component to humans namely their language, culture and mind. This non-physical component, however, is symbolic and not necessarily spirit-like or transcendent. Neo-dualism as developed by Logan and Schumann (2005) dispenses with or is agnostic with respect to the notion of soul, spirit or God but assumes that human behavior consists of both a physical and a non-physical component. The non-physical component is not necessarily spiritual but rather is conceptual or symbolic. The concepts of zero, energy, numbers, force, life, morality, democracy, liberty, and marriage, for example, do not have a physical or material instantiation. They are non-material products of the human mind and they are without extension.

Neo-duality makes an explicit distinction between purely material phenomena whether they are abiotic or biotic and non-material phenomena associated with human thought namely, ideas, symbols, language, culture, and the concepts that go into creating science, technology, governance and economics, artistic creations and music. In the neo-dualistic approach of Logan and Schumann (2005) all phenomena belong to one or the other of two different domains: the physiosphere and the symbolosphere. The physiosphere is simply the material world consisting of both living and non-living matter and corresponds exactly to Descarte's res extensa the domain of things with extension. The symbolosphere consists of the human mind and all the symbolic products of the mind and corresponds to Descarte's res cogitans minus the notions of God, the soul and spirit. The symbolosphere like Descarte's res cogitans has no extension or physicality.

In our neo-dualistic model the human brain and the mind are seen as distinct entities with the brain belonging to the physiosphere and the mind to the symbolosphere. This model of neo-dualism grew out of Schumann's (2003a & b) notion of the symbolosphere and Logan's notion of the Extended Mind (1997, 2000 & 2007), which posits that the mind is the product of the human brain plus verbal language. Neo-dualism represents a weak form of dualism as contrasted with the strong dualism of Descarte.

Clayton (2004, v) has suggested that dualism and emergence are in conflict, "Emergence ... represents a third option in the debate and one that is preferable to both of its two main competitors," dualism and physicalism. If Clayton restricts his notion of dualism to the Cartesian one that posits the existence of a spiritual substance to explain the existence of God and the human soul then emergence and dualism are in direct conflict. If however one considers the neo-dualist position as developed by Logan and Schumann (2005) then the conflict disappears and the position of emergence and neo-dualism, as we will demonstrate, are perfectly compatible.

The focus of this chapter is to articulate this notion of weak dualism or neo-dualism in light of propagating organization as described in *Investigations* (Kauffman 2000), in POE and in the non-material emergence and the propagation of organization as described in the last chapter. In carrying out this analysis we will carefully make the distinction between material gene-based propagating organization in the biosphere as described in POE and non-material, extra-somatic, meme-based propagating organization in the symbolosphere of human language and culture.

To conclude this introductory section we emphasize that neo-duality embraces strong emergence but make a clear distinction between the materiality of the biosphere and the symbolic non-material nature of human language, conceptual thought and culture.

Cartesian Dualism and Neo-dualism: A Comparison

Descartes' dualism has fallen into disfavor within the scientific community and large parts of the philosophical community that embrace the scientific method. The reason is that Descartes introduces into his philosophical system entities that cannot be empirically probed such as soul, spirit and the Deity and which properly belong to the realm of belief and theology. "Strongly dualist theories of human nature, and in particular substantival theories of the soul, have become problematic in an age of science (Clayton 2004, 124)." The position of most scientists and philosophers of science with respect to these categories introduced by Descartes into his philosophy is one of agnosticism in their pursuit of science or their understanding of how science operates. On the personal level scientists and philosophers of science range from true believers to agnostics to atheists and even to belligerent atheists who feel the need to belittle theists.

In formulating res extensa, the domain of the material, and res cogitans, the domain of the non-material or conceptual, however, Descartes made an important distinction between the material and non-material domains of this world that have extremely important implications for biology, anthropology, sociology, economics, political science, and media ecology. With the exception of biology all of the disciplines listed deal almost exclusively with res cogitans; whereas human biology deals with a mixture of the two as is the case with both evolutionary biology and biosemiosis where information in both material and non-material formats influence the evolution, development and the survivability of humans.

The Extended Mind

Our definition of res cogitans that we have just given is incomplete, however, unless we describe exactly what we mean by the human mind, which as has been posited in the Extended Mind model (Logan 2000, 2006a and 2007) is different than the human brain. The mind can be thought of as the processor of symbolic thought whereas the brain is a percept processor and mind = brain + language. Speech and the human mind emerged simultaneously as the bifurcation from percepts to concepts and a response to the chaos associated with the information overload that resulted from the increased complexity in hominid life. Verbal language and abstract conceptual thinking emerged together at exactly the same point of time as a bifurcation from alingual communication skills and the concrete percept-based thinking of pre-lingual hominids to verbal language and conceptual thought (Logan 2000, 2006a and 2007).

Res Cogitans or the Symbolosphere

Res extensa or the physiosphere consists of the material world and hence everything that has extension and is made of stuff, ultimately atoms or elementary particles or if one wants to go to an even deeper level, leptons and quarks (and some would claim strings but there is not one shred of empirical evidence for these). Res cogitans or the symbolosphere is everything else. It is the non-material world or the symbolosphere and consists of the human mind and all of the mind's concepts and analytic tools such as language, culture, science, technology, laws and economics. None of these elements of res cogitans or the symbolosphere have extension or are composed of material components. They emerged from the behavior and interactions of the human animal and they have a downward causation on the humans from which they emerged. The difference in the emergence of res cogitans or the symbolosphere from other forms of emergence like the emergence of the biosphere from organic chemistry and hence atoms is that living organisms are composed of atoms but the elements of res cogitans are not composed of anything material but rather are the products of human thought and behavior.

The one characteristic that unites all of the elements of res cogitans or the symbolosphere is that they are all symbolic. Terence Deacon described humankind as the symbolic species and res cogitans or the symbolosphere is the set of symbolic elements that comprise the behaviors of the symbolic species. John Schumann and N. Lee have a very succinct way of describing the relationship between the abstract, non-material, non-extensive element of language and the material extensive human brain from which language emerged and in which language operates in a downward causal manner. Schumann (2003) suggests that the words and grammar of language emerge as a complex adaptive system as a result of the communicative interactions of hominids. Language as a consequence is a cultural artifact; it "is neither of the brain nor in the brain (Lee and Schumann 2003)." Its organization does not propagate biologically but rather culturally and "exists as a cultural artifact or technology between and among brains (ibid.)."

Language is an artifact that is non-extensive and non-material and hence is not part of the material biosphere but rather is part of the symbolosphere. The symbolosphere includes all forms of symbolic communication including spoken and written language, mathematics, science, technology, computing, the Internet, laws, economic systems, music and the arts. Each of the elements of the symbolosphere propagates its organization just as living organisms do. The difference is that the mechanism for replication for living organisms is chemically based through DNA whereas the replication of the linguistic and cultural elements of the symbolosphere is through memes. It is also the case that just as living organisms evolve through the mechanism of descent, modification and selection the same is true of the elements of the symbolosphere. The descent occurs each time a meme is transmitted from one mind to another. A modification can take place in the mind of the recipient of the meme if he or she so chooses. And the selection process occurs when other human minds decide whether or not to adopt the new or modified meme.

Culture

Culture is an important adaptive mental tool that is more or less unique to humans whereby the learning of previous generations are passed on to the next generation through communication and social interactions. Culture like language is another symbolic activity which is abstract, non-material and non-extensive.

Geertz (1973, 8) defines culture in symbolic terms as does Durham (1991, 8–9) when he wrote,

> the new consensus in anthropology regards culture as a system of symbolically encoded conceptual phenomena that are socially and historically transmitted within and between populations. As Keesing has pointed out, this view contrasts markedly with earlier conceptualizations of culture as adaptive behavioral systems, for which human populations maintain themselves in local environment.

Culture includes technology, economics, governance and science each of which is symbol based. Culture is a form of propagating organization that evolves like living organisms by descent, modification and selection as described above. Culture represents the way in which a society organizes its material life of food, shelter, clothing, protection, etc. This organization is symbol-based but has a downward causative effect on the material artifacts of society and the behavior of its members.

Economics and Governance

Economics and governance are another element of culture that organizes human interactions and creates social cohesion. This form of propagating organization is symbolic as is pointed out by Johnson and Earle (1987, 322):

> To sustain economic integration beyond the capacity of the biological bonds that underpin the familistic group, it is necessary to extend the individual's sense of 'self-interest' to broader social units. This extension of self is based on symbols.

Economics and governance although they are symbolic and non-material they still have a downward causative effect on the human agents in which these forms of organization reside. The ways of making a living and organizing society descend from one generation to another but are subject to modification as environmental conditions change or as individuals in a society innovate. Those modifications, which better support the society, are then selected completing the process of Darwinian evolution of descent, modification and selection.

Technology

Technology is another element of culture, which at first blush seems to be material. Actually technology is conceptual and symbolic and represents the way in which materials are organized through downward causality to achieve functionality. Technology is therefore a form of propagating organization that also evolves like living organisms by descent, modification and selection. All technologies are derived from or descend from some earlier tool. The very first human tools were derived from found objects, as is the case with primates that make their tools from found objects. Tools descend from generation to generation. The inventor or designer of a new technology is the source of modification of some older tool or combination of tools. Finally, the users who opt or select to use the technology complete the evolutionary cycle of evolution. Those tools that are functional and easily and comfortably deployed are selected.

Science

Science is the final element of culture that we will analyze. Science is basically a non-material symbolic methodology for describing nature. Thomas Kuhn (1972) in *The Structure of Scientific Revolutions* described the descent and propagation of the organization of science through the articulation of normal science. The period of revolutionary science is the period of modification of existing theories by which new scientific laws and descriptions of nature emerge. The empirical verification of scientific hypotheses completes the selection component of the evolutionary cycle. The downward causation of science operates on the other elements of culture such as economics, governance and technology.

Conclusion

We have shown that all the elements of language and culture including explicitly economics, governance, technology and science are all non-material, emergent and represent propagating organization justifying the neo-duality approach to understanding reality and the compatibility of strong emergence and neo-duality or weak duality. One of our objectives was to answer the question: Is information material, a form of energy or is it just a pattern? I believe that we have shown that biotic or instructional information is material and that the information contained in language and culture is symbolic and is a pattern. Biotic information instantiated in DNA, RNA and proteins can be construed as a form of chemical energy for whatever that is worth.

In closing this chapter I wish to acknowledge that the ideas presented here were stimulated by John Schumann, my co-author of (Logan and Schumann 2005) where we first formulated the notion of neo-duality.

21er-Haus. Ground Floor. Belvedere, Vienna by Alfred Weidinger http://www.flickr.com/photos/a-weidinger/6234742133/

Chapter Seven

The Four Spheres of Influence on Human Existence

We have already indicated that with respect to our neo-duality model of reality described in the last chapter that phenomena can be assigned to either the physiosphere of material reality or the non-material conceptual reality of the symbolosphere.

The physiosphere can be divided into an abiotic part that can be described purely by physics and the biosphere of living organisms that arises from the physiosphere as an emergent phenomenon. The symbolosphere of non-material phenomena arises as emergent symbolic entities from that part of the biosphere that includes human life.

In this chapter we will return to the roots of the term information in the English language, which originally denoted giving the human mind form and therefore examine the impact of information on human existence. There are three forms of information that impact the human condition. They are the genetic information of our DNA, the perceptual information that we detect with our senses and the conceptual information that we process with our minds, which I remind the reader is the product of the brain and language. Human existence is therefore moderated by two spheres of influence and organization in which the transmission of information and the propagation of organization is key. The two spheres are the biosphere and the symbolosphere.

The biosphere is part of the physiosphere or res extensa, which also incorporates the abiotic material world. The abiotic part of the physiosphere also influences human condition in that we are subject to the laws of physics. We are acted upon by gravity, we are subject to the laws of thermodynamics and we are affected by electromagnetic radiation. But we do not consider the laws of physics that rule the abiotic physiosphere as information or a form of organization that is propagated but rather as the ground in which the operations of the others spheres take place. The laws of physics are pervasive—they are not propagated as organization as is the case with the information in all the other spheres under consideration. The information contained in the other spheres is localized whereas the laws of physics are ubiquitous.

The symbolosphere consists of the mind (as opposed to the brain), language; culture; political economy or the econosphere.; and technology or the technosphere. The technosphere and the econosphere are actually subsets of the symbolosphere, but it suits our purposes to treat these two spheres separately and to consider language and culture in general under the heading of the symbolosphere. The reason for this division is that the technosphere and the econosphere each have unique mechanisms for the propagation their organization and their evolution and they each have a downward causation on the material world of the physiosphere.

The technosphere consists of all the concepts that go into the organization of human tools and technologies. The technosphere is purely symbolic and does not include the materiality of the physical tools but rather the symbolic concepts that are used to organize elements of the physiosphere into the physical tools and technologies that we humans make use of.

The econosphere consists of all the concepts for economic and governmental activities such as businesses ranging from corporations to sole proprietorships, NGOs and government agencies ranging from the offices of the heads of states, parliaments, judiciaries and various administrative bodies such as ministries and departments. The elements of the ecosphere that we are considering are conceptual and represent the organization of these organizations (pun intended) and not their actual day-to-day operations in the physiosphere. The term econosphere was first coined by Kenneth Boulding (1966).

> We can think of the world economy or "econosphere" as a subset of the "world set," which is the set of all objects of possible discourse in the world. We then think of the state of the econosphere at any one moment as being the total capital stock, that is, the set of all objects, people, organizations, and so on, which are interesting from the point of view of the system of exchange. This total stock of capital is clearly an open system in the sense that it has inputs and outputs, inputs

> being production, which adds to the capital stock, outputs being consumption, which subtracts from it.

Our definition differs from that of Boulding as his econosphere contains the actual physical elements of the economy, which is why it is open to energy and matter as well as information. The econosphere that we are considering is purely symbolic consisting of the patterns for the economy and not the patterns of the economy paralleling the way Geertz defined culture as a symbolic entity. Our econosphere is therefore open only to information and not to energy and matter. The economic organizations that compose the econosphere are technologies in a certain sense in that they are systems and processes for exchange. It is useful, however, for the purposes of our analysis to consider them separately from technology.

The elements of the symbolosphere arise as emergent phenomena from and are nested in the biosphere. The biosphere, on the other hand, arises as an emergent phenomenon from and is nested in the abiotic physiosphere. The upper spheres arise from the sphere just below them as emergent phenomena and act through downward causation on all the spheres below them. Not only does one have to understand the biotic information stored in human DNA, RNA and the proteins but also all of the symbiotic organisms from both the biosphere and the symbolosphere that live within the human organism plus all the other organisms with which humans interact. As our focus is on information we will confine our study to the obligate symbiotic organisms of the symbolosphere such as language, culture, technology, political economy and ignore the biotic symbionts that also live within us.

The motivation for the division into the four spheres of influence on human existence is that we want to study their information content, the way in which they evolve, and their openness to information flows (and in the case of the biosphere the openness to both energy and information flows), which differs from sphere to sphere. We also wish to examine the agency within each of the spheres and symbiosis of those agents with each other and with their human hosts, which also differs from sphere to sphere. In some case we will be reviewing some of the ground we have already covered and in other instances such as our discussion of agency and symbiosis we will be covering new ground. Another reason for the division of the influences into the four spheres is a heuristic one since it facilitates the comparison of the different forms of information that influence us humans. We hope in this way to make a grand synthesis of our attempt to answer the question what is information before we take up some examples in chapters 8 through 10.

Information Content, Evolution, Agency, Openness and Symbiosis

As was argued in Chapter 5 the elements of the symbolosphere including languages, cultures, technologies and economic entities can be treated as living organisms, albeit parasites or obligate symbionts, that propagate their organization. Living organisms contain biotic information in terms of the organization they propagate. They evolve by the Darwinian process of descent, modification and natural selection. They also have agency in that they are autonomous agents that act on their own behalf. They are open to matter, energy and information and they enter into symbiotic relationships with each other like the example of the fungus and the plant discussed in Chapter Five. All of these properties of autonomous living organisms that populate the biosphere as we will show are also true of the elements of the symbolosphere. They too contain information, evolve, have agency, are open to information and enter into symbiotic relationships with their human hosts and with each other. In this chapter we will consider and compare from the point of view of the transmission of information and the propagation of organization the elements of the four spheres of influence on human existence.

We are particularly interested in comparing within the biosphere, the symbolosphere, the technosphere and the econosphere the following five properties of their constituents:

1. the nature of the information they contain,
2. the way in which they evolve,
3. the nature of their agency,
4. their openness to matter, energy and information, and
5. the symbiotic relationships they enter into.

The nature of the symbolic constituents of the symbolosphere, technosphere and econosphere parallels the nature of biotic living organisms. They contain information in the organization of their structures. They evolve through descent, modification and selection the mechanism that Darwin identified for living organisms. They are not autonomous agents that are able to source free energy on their own. As obligate symbionts they depend on their human hosts for their source of energy but they also enhance their hosts capability for exploiting sources of free energy and hence their relationship with their human hosts is symbiotic rather than parasitic. As a result we will consider only their openness to information as their openness to energy is through their human hosts and there is no openness to matter as they are non-material. Although they are not autonomous they are agents in the sense that they have causal effects and they act in their own self interest in that they insure the propagation of their organization. They are symbiotic in the sense that the various symbolic elements of human culture in the same sphere and between spheres reinforce each other, work together as coherent parts of human existence and co-evolve.

Biosphere

> *Biology is not a physical science but a semiotic science*
> *—Terrence Deacon*

A living organism that populates the biosphere is an autonomous agent that is also

1. a heat engine converting free energy into work;
2. a factory for fabricating complex biomolecules from the raw materials of its environment;
3. an information processing device converting external signals into appropriate internal actions and subsequently appropriate interactions with its external environment; they convert the raw data of environmental information into complex behaviors that allows them to source the energy and raw materials they need to propagate their organization and to avoid the toxins and predators that might terminate their existence. While it has been argued that living organisms are computing devices they should not be confused with human manufactured computers, which are devices, that process symbolic information. The computing or information processing of living organisms is not symbolic but rather involves the processing of information that emanates from material substances;
4. a medium of communication generating appropriate messages to its conspecifics and other organisms. They also process environmental data to enter into social and symbiotic relationships with both conspecifics and other species and as such they may also be considered as a medium of communication.

Each of these activities is necessary for the living organism to sustain itself and its species by carrying out its metabolism and its replication or in other words propagate its organization. Each of these activities in one way or another involves **information**. Activities 1) and 2) require the constraints we identified as instructional information when we reviewed POE in Chapter 2. Activities 3) and 4) involve the flow of information back and forth between the organism and its environment. There are two types of information associated with biotic systems—internally inherited information needed for activities 1) and 2) by virtue of the organism's organization, which it propagates by replication and growth and the information that flows through the organism as it carries out activities 3) and 4). The flow of information from the environment to the living organism is processed by the organism resulting in another flow of information back to the environment in terms of the organism's interaction with its physical environment and/or the other living organisms with which it interacts whether they are conspecifics or members of another species.

There is a third kind of information, namely human generated symbolic information, which we deal with as part of the symbolosphere. The internal inherited information content consists of the organization that the living organism propagates. An example of this biotic or instructional information are the constraints that allow the organism to convert free energy into the work required for their metabolism and replication. Biotic information also includes their DNA, RNA and proteins that guide the development and growth of the organism.

> As part of the propagating organization within living cells, the cell operates as an information processing unit, receiving information from its environment, propagating that information through complex molecular networks, and using the information stored in its DNA and cell-molecular systems to mount the appropriate response. Indeed, biology is acquiring many characteristics of an information science (Hood and Galas 2003)."

The above quote describes the information processing activities of a cell but there is also the flow of information between cells in multi-cellular creatures. In simple animals without a brain there is the gathering of information by receptors and/or sense organs, which leads directly to the action of the motor system. In plants information flows give rise to phenomena such as heliotropism. Finally with animals with brains there is the flow of information to the brain where it is processed and then results in signals to different parts of the body such as the motor system.

Among the four functions of living organisms that we have identified, namely informatics, energetics, bio-fabrication and communications there are many linkages. First of all the information contained in their organizational structures permit 1. the conversion of free energy into work, 2. the bio-fabrication of bio-molecules, 3. the processing of information from the outside world and its transmission to different parts of its structure and, finally, 4. its communication with others. The chemical reactions building the biomolecules that make life possible are endothermic and hence require the living organism to harness free energy from its environment and convert it into work or stored energy for times that are lean. As to the organisms ability to communicate this must arise from its ability to process information and select the appropriate message to communicate as well as having energy to carry out this activity. Communication, on the other hand, is essential for sex and hence the replication of organization as well as cooperative ventures that enhance the acquisitions of energy. And finally biofabrication is an essential part of the organism's metabolism and replication.

In view of this interlocking of energetics and informatics I am inclined to added to Kauffman's definition of a living organism or autonomous **agent** as an autocatalytic system carrying out at least one thermodynamic work cycle the condition that it is also capable of carrying out the processing of information derived from its environment such as the location of free energy and the presence of dangers such as predators or toxins.

Since living organisms sustain themselves by their ability to source free energy and convert it into work they are **open** to energy. They are also open to matter, which they require along with energy for growth and replication. Finally they are also open to information through their receptors that are part of their organization and help them to source free energy and to avoid toxins and/or predators.

We have claimed Darwin's model of descent, modification and selection with which he described the **evolution** of living organisms also provides a model that describes the evolution of the various elements of the symbolosphere.

Another important characteristic of living organisms is that they enter into **symbiotic** relationships with each other. Symbiosis literally means living together. All biotic agents from the simplest bacteria to the most complex plants and animals including humans enter into symbiotic relationships with other organisms. There are many examples. We humans could not survive without the many other organisms that live within us. Fungi that live on the roots of plants fixing nitrogen in an exchange for food that we have referred to earlier represent a classic case of symbiosis. Symbiotic relationships develop through the communication of information.

The one form of symbiosis that had the greatest impact on the biosphere and allowed life to evolve beyond single cell bacteria, i.e. prokaryotes, and gave rise to the animal and plant kingdoms was the emergence of eukaryotes. Eukaryotic cells emerged from the symbiosis of two bacteria. The bacteria that contributed to the nucleus of the eukaryote cell combined with mitochondria in the case of animals or chloroplasts in the case of plants. It is surmised that the first eukaryotes arose when one bacterium, for example a mitochondria, penetrated the wall of another bacteria to create a symbiotic relationship of two organisms that were originally prokaryotes. These two formerly autonomous organisms lived within the walls of a single cell to become a single eukaryote organism more complex than its two original prokaryote components. The birth of eukaryotes represents an example of emergence because the eukaryote's properties cannot be reduced to, derived from or predicted from the properties of the two prokaryotes of which it is composed. Even before the emergence of eukaryotes bacteria formed cooperative symbiotic networks or relationships in which one bacterium provided a function or service for another bacterium in exchange for food. It was surmised by Lynn Margulis (1970) that in the course of one of these symbiotic relationships one of the bacteria instead of lingering in the neighborhood of its partner actually entered into the cell of the other and thereby surrendered its autonomy. The resulting eukaryote cell then became more complex with one of the bacteria forming the nucleus and the other the organelle of mitochondria, the energy engine of the cell.

The Symbolosphere: Mind, Language and Culture

Homo sapiens emerged from biosphere simultaneously with the emergence of the language and the human mind that bifurcated from the brain with language and culture (Logan 2007) as was discussed in Chapter 3 and 4. These developments represent a form of punctuated equilibrium where natural selection can result in sudden discontinuous changes on time scales that are relatively brief on the geological time scale and correspond to speciation events, followed by longer periods of less dramatic change (Eldredge and Gould 1972). These three elements of language, mind and culture form the symbolosphere (Logan and Schumann 2005) as discussed in Chapter 5.

Mind: The information content of the mind of each individual are the languages and all of the concepts that they possess. One cannot speak of the **evolution** of the mind because evolution requires descent, modification and selection. An individuals mind does not survive their death and therefore there is no process of descent as is the case with language and culture. As to the other categories we are examining in this chapter one can make the case that each individual mind has **agency**. Some who embrace a radical form of reductionism might challenge this notion claiming that human action is totally determined and that free will is only an illusion. We will not enter into this debate as this study is based on the validity of an emergentist point of view, the validity of which cannot be proven according to the ideas developed in the next chapter. Besides our objective is to develop a better understanding of the nature of information and not to debate the merits of free will versus determinism. As to other categories we are examining in this chapter it is pretty obvious that individual minds are **open** to information and that they enter in **symbiotic** relationships with other minds, as well as language, culture, technology, economics and governance.

Language: Words and language gave rise to the symbolosphere as words are abstract symbols that represent something other than the physical sign that is the word. We begin our analysis by studying the five properties of language or words. Words are both a medium for representing, communicating and expressing information and a form of **information** in themselves. Words and language are a pure medium for the representation of information and like the light bulb is, according to McLuhan (1964, 8), a medium without content. The user (or speaker) provides the content or as McLuhan once said "the user is the content."

The representation of information with words or language needs little justification but perhaps a word or two for the claim that words themselves are a form of information is in order. As was argued in Chapter 3 words act as strange attractors uniting all of the percepts associated with the concept that the word represents. Words therefore carry with them the information of all the percepts with which they are associated plus all of the ways in which the words, as part of a semantic web, have been used.

Because they carry the information as a result of all of their association words have **agency** in that they carry and assert (or insert) additional meanings that the speaker might not have intended when using them. Poets are adept at making use of the multiple meanings of words allowing the agency of the words they choose to use to enrich their poetry.

Words are **symbiotic** in that they live together and work with each other. A word isolated by it has no meaning. Words give meaning to each other through the syntax of the language and within the semantic web in which they exist. They change each other's meaning. They sometimes co-exist together in a single new word as house and boat in houseboat or in expressions like "by and large" or in compound words like steamboat, steamship, and airplane.

The meaning of words **evolves**. The word fair for Shakespeare meant beautiful or wonderful whereas today it means average. The word awful originally meant something that filled one with awe but now means something that is unpleasant.

Language as a medium for the representation and transmission of information is naturally **open** to information. A language is also open to the information in the sense that it is open to other languages through the use of loan words and their participation in a sprachbund, a union of languages that have certain similarities because of geographic proximity. Words are open to information in that they are part of and form a semantic web and they are open to each other through the grammar or syntax of the language. Language is also open to new information generated by the experiences of their users. New words are invented as novel experiences emerge such as the invention of new technologies or new social, political or economic situations. Existing words take on multiple meanings like the word "cool".

If we consider the language of each individual as an organism and a language like English or French as a species of all the individuals who can communicate with each other through that language then we see that the languages of individuals live together with the languages of other individuals, i.e. they are symbiotic. The species languages of English and French also interact as English speakers use French words such as chauffeur or RSVP and French speakers use Englishisms such a le weekend or le pullover.

Culture: Culture is a symbolic form of **information** that is transmitted from generation to generation. It is according to Geertz (1973) "patterns for behavior". Culture also serves as a medium through which information is generated and conveyed. As described in Chapter 4, cultures evolved in such a way so as to be easily learnt so that they can propagate their organization. The ability to propagate their organization and thereby preserve themselves reflects the **agency** of cultures. So does the fact that societal cultures often act in their own self-interest at the expense of their hosts as is the case of an imperial culture, which requires the sacrifice of its citizens when calling upon them to put themselves in harms way during military operations. Cultures change or **evolve** to improve the chances of the survival of their hosts and hence themselves as environmental and/or social conditions change. They are able to do this, as they are **open** to information. As described in Chapter 4 the organisms of culture are those that belong to individuals, which interact with each other to form a society. The cultural organisms of individuals are therefore **symbiotic** in the sense they live together and by so doing they create the *culture* of their society. We will examine in detail two important elements of culture, namely, the technosphere and the econosphere in the remainder of this chapter. Then in Chapters 8 through 10 we will study three other elements of human culture, respectively, science (8), the book (9) and the creative arts (10).

The Technosphere

The technologies including products, services, processes and systems that have been invented by humankind belong to a space akin to the biosphere that is commonly known as the technosphere. The technosphere as we utilize the concept in this study consists of the abstract symbolic concepts that go into creating and using tools or technologies and not the actual physical tools or technologies themselves. These concepts, which are the patterns (ala Geertz) for the manufacture and utilization of tools, therefore form the **information** content of the technosphere.

Technologies, unlike living organisms that are able to sustain themselves by carrying out thermodynamic work cycles, are not autonomous agents. Rather technologies are obligate symbionts that depend on their human hosts for their inception, manufacture and the energy for their operation and their action on the environment and their human hosts. Technologies are therefore not autonomous and their **agency** is on the whole directed by the intentions of their human users. The question therefore arises as to what extent can we claim that technologies really possess agency. We begin addressing this question by asking what is an agent and what we mean when we attribute agency to an object or a process. A dictionary definition of an agent is one who acts or causes things to happen. This is an obvious trait of a living organism that acts on its environment exploiting its resources in terms of raw materials and free energy to propagate its organization and as a consequence causes changes to its environment. But, what about technologies? Let us consider a hammer. A hammer is an inert object that cannot of its own accord pound a nail into wood. It can only effect change through the agency of its human user. It is through its use by a human user that the hammer and its user together have agency. But looked at through a McLuhanesque perspective the hammer may be regarded as an extension of its user and therefore the hammer partakes of the agency of its user.

It can also be argued that technologies possess agency by virtue of the fact that they also act upon their human hosts and the environment in ways that are independent of the intention of their users. McLuhan through his aphorism "the medium is the message" which includes all forms of technology and his observation that the effects of media and technologies are often counter-intuitive and unintended identified a certain level of agency for technologies and media. While strictly speaking technologies or media do not literally initiate their actions they do so metaphorically because part of their impacts or actions are totally unrelated to the intentions of their users. It was not the intention of the automobile to create the suburbs but suburbs emerged nevertheless because of the automobile. One can therefore claim it was the agency of the automobile, which created the suburbs.

Like biological living organisms technologies **evolve** through the Darwinian mechanism of descent, modification and selection as pointed out by Basalla (1988) as was discussed in chapter 5. The mechanism of descent applies to technologies because all tools start as a modification of a former tool with the very first original tools being found tools. The modification of the starting tools is done by the inventor and sometimes involves the convergence of two tools as was the case of the automobile or horseless carriage as it was first called, a marriage of the carriage and the motor.

The fact that two tools can combine to form a third tool illustrates that tools are **open** systems. The earliest example is the construction of the axe with a handle, which is the combination of the hand axe, which became the head of the axe and the lever in the form of a stick, which became the handle of the axe. After the invention of the steam engine the steam engine replaced the water wheel and the windmill and combined with the mechanical tools that were powered by running water and wind. These examples also illustrate how different tools like living organisms can enter into **symbiotic** relationships with each other in which the technologies support each other. The example of the automobile emerging from the horse drawn carriage and the gasoline engine is an example of techno-symbiosis that parallels the emergence of eukaryote cells where the carriage plays the role of the cell with a nucleus and the engine plays the role of the mitochondria. The symbiosis of technologies (techno-symbiosis) can takes place with stand alone technologies supporting each other as is the case with the iPod, iTunes, personal computers and the World Wide Web. Users are able to download songs from the Web-based application iTunes onto their personal computer and from there upload the song on to their iPod. Apple Inc. created iTunes and the iPod to work together symbiotically taking advantage of the Web and notebooks.

The iPhone represents the symbiosis of a group of technologies that include the Web, the cell phone, the camera, the iPod, and the touch sensitive screen that combine to create this device. The iPhone is an emergent phenomenon in that it has properties in addition to those of its components that cannot be derived from, predicted from or reduced to the properties of its components. The success of a new technology depends on what techno-symbiotic relationships it can form with other technologies that support its success. For example the technology of the automobile requires the technology of roads and a distribution system for gasoline. Other examples of techno-symbiotic relationships include:

- cultivation and irrigation;
- writing and paper;
- the movable type printing press, the alphabet, and paper;
- the book and the printing press;
- the steam engine and mechanical devices such as the locomotive and the steamship;
- the skyscraper and the elevator;
- the automobile and the highway;
- the electrification of mechanical devices, such as electric motors and the phonograph player or any of a variety of electric kitchen appliances;
- the mainframe computer and programming languages; and
- the personal computer and software applications.

The analogy between living biotic organisms and technologies is fairly compelling given that they each have agency, evolve new forms, have information content, are open to information and enter into symbiotic relationships with each other. The only things lacking is that they depend on their human hosts for the energy of their operation and they are unable to reproduce themselves but require human intervention for their reproduction. Techno-organisms are therefore not autonomous agents but rather they enter into symbiotic relationships with their human hosts and may be regarded as obligate symbionts.

There is an interesting spin on the notion of technologies behaving like organisms in the sense that when they combine or cross-pollinate with each other—it is as though they are mating. Thus when the carriage of the horse and carriage mated with the motor the automobile was born. Or when the boat and the steam engine mated the steamboat was born.

In the 20th century computer scientists, cyberneticians, information theorists and artificial intelligence (AI) experts made use of the analogy or metaphor of the computer or Turing machine for creating their physicalist's models of life and intelligence just as Newtonian physics gave rise in the 18th century to mechanical models of life and intelligence. More recently with the development of emergentist self-organizing models of life and intelligence in the work of Kauffman (2000) and Maturana and Varela (1992) the metaphor of Turing machines is being superceded with biological models. This is the tack that I have taken in this study which I believe provides fresh insights into the nature of human behavior such as the invention and use of technology. If the focus of 20th century models was on computing I believe twenty-first century models will focus on biology and perhaps other emergent phenomenon like culture, language and the mind.

Disruptive technologies as saltations or examples of punctuated equilibrium

To support our notion that the evolution of technology is similar to the evolution of living organisms in the biosphere we will consider examples of disruptive technologies, which in the technosphere function as forms of punctuated equilibrium as developed by (Eldredge and Gould 1972) and was described earlier in this chapter. Like punctuated equilibrium in the biosphere disruptive technologies represent sudden discontinuous changes in the array of human tools in time scales that are relatively brief on the time scale of the technosphere and correspond to a new technological era in which a number of new technologies arise taking advantage of the new disruptive technology while at the same time some older technologies become obsolete or take on less important functions.

We identify these technologies as disruptive because they led to major shifts in the development of other technologies and they brought about major social, economic and political change. Examples of disruptive technologies, which ushered in associated changes include:

- the first stone tools and the control of fire over one million years ago from which mimetic communication and culture emerged,

- the explosion of cultural artifacts or technological innovations in tool making circa 100,000 to 50,000 BCE, which many believe, corresponded to the same period that human language and symbolic representation such as art emerged (Bickerton 1998, 354–5; Crow 2002, 93; Dunbar 1998, 105; Logan 2007),
- writing and mathematical notation circa 3,000 BCE associated with the rise of city-states and civilizations with written laws and religious institutions,
- science circa 2,000 BCE and then modern science circa 1500 AD,
- the alphabet circa 1500 BCE associated with monotheism, philosophy, and deductive logic and led to the discovery of zero,
- zero and the place number system circa 200 BCE associated with algebra
- mechanical devices such as heavy plow, animal harnesses, wind mills, water wheels circa 1000–1400 AD associated with the rise of the bourgeoisie and modern cities and led to the printing press,
- movable type printing press circa 1450 AD associated with the Renaissance, the Reformation and eventually universal education and provided a model for mass production,
- steam engine associated with the industrial revolution circa 18th Century,
- electricity associated with electric mass media circa 19th to mid 20th century,
- computing in the last half of the 20th century and also associated with automation and robotics,
- Internet and World Wide Web from 1980 to present associated with Web 2.0 economics.

Each of these forms of disruptive technology is a perfect example of punctuated equilibrium. These discontinuities in the evolution of technology illustrates Prigogine's theory that far from equilibrium new levels of order can suddenly emerge as a bifurcation from a chaotic non-linear dynamic system which is the nature of human culture
(Prigogine and Stengers 1984 & Prigogine 1997).

Exaptations

Another parallel of evolution in the biosphere and the technosphere are the presence of exaptations or spandrels. In St. Mark's Cathedral in Venice spandrels are architectural structures that are integral to the support the building. They were decorated with images of the evangelists and are an integral part of the decoration of the interior of the church. Gould used the metaphor of the spandrel to explain the phenomenon of exaptations in the evolution of biological organisms:

> Under the spandrel principle, you can have a structure that is fit, that works well, that is apt, but was not built by natural selection for its current utility. It may not have been built by natural selection at all. The spandrels are architectural by-products. They were not built by natural selection, but they are used in a wonderful way—to house the evangelists. But you can't say they were adapted to house evangelists; they weren't. That's why Elizabeth Vrba and I developed the term 'exaptation'. Exaptations are useful structures by virtue of having been co-opted—that's the 'ex-apt'—they're apt because of what they are for other reasons. They were not built by natural selection for their current role (Gould 1996, 59).

An example of biological exaptations is the wings of insects, which originally served as cooling devices but were exapted for flight. The same is true of dinosaur wings, which were originally upper limbs that were used to scoop up prey more efficiently and were exapted into devices for flight. Another example was the swim bladder that fish used to regulate the depth to which they could descend by changing the mixture of air and water the bladder held. This device exapted into lungs and resulted in the emergence of land animals.

Exaptation play an important role in the design of technology. Examples include the way in which three dimensional clay tokens discussed in Chapter 3 used for accounting in Sumer became exapted into two dimensional signs for agricultural commodities and numerical quantities that eventually evolved into writing and numerical notation. The Newcomb steam engine designed to pump water out of coalmines was exapted by James Watt to harness steam for locomotion and the operation of factories that had previously been powered by water wheels and windmills. The Gutenberg movable type press is another example of an exaptation of the wooden block print system of text which itself was an exaptation of textile printing. Exaptation is powerful tool in the design of new technology.

An Alternative to the Darwinian Evolution of Technology

We have to this point suggested a Darwinian model of descent, modification and selection for the evolution of technology and even drew analogies with punctuated equilibrium and exaptations. There is one difference, however, between the evolution of biological organisms and technology and that is in the modification step. In biology the modifications are random and unintentional whereas for technology the modifications are intentional and chosen by the designer of the new technology. As a result Olesen (2008) has suggested that perhaps the model of the Darwinian evolution has to be modified somewhat due to the role of the designer or innovator of the new technology. He wrote,

> Neither a Darwinian random selection process nor a completely rationalized, planned Lamarckian-like process is a proper description of how media develop. We need a combination describing what perhaps could be called a new kind of evolution, alternative to the biological. Still, a fundamental question remains: where do the purposeful creations of the designer stop and the general mechanisms of overall media development take over?

Technological Innovation, Design and Emergence

While Olesen was correct to point out the difference between bio- and techno-evolution, there still exists a parallel of the two processes. Van Alstyne and Logan (2006) in a study of industrial design at Strategic Innovation Lab (sLab) at OCAD University have argued that the very act of designing an innovation involves a process of emergence similar to that of the emergence in the biosphere. They discovered the surprising and counterintuitive truth that the design process, in and of itself, is not always on the forefront of innovation. Design is a necessary but not a sufficient condition for the success of new products and services. They proposed that design must harness the process of emergence; because it is only through the bottom-up and massively iterative unfolding of emergence that new and improved products and services can be successfully refined, introduced and diffused into the marketplace.

They suggested the following parallels between the emergent design of technology and biological emergence:

- propagation of organization toward a goal or purpose,
- involvement of selection,
- development of differentiation from generality or an increase in complexity,
- morphogenesis or the birth of new forms.

They also identified the following differences between human design and emergence in nature:

- intentionality of the technology designer versus autonomy of massively multiple biological agents
- technology design is cognitive, conceptual, top down, controlling

versus biological emergence, which is just the opposite, a-cognitive, a-conceptual, bottom up, non-controlling
- fixing relationships versus maintaining relationships, and
- setting constraints versus exploring constraints.

Emergence as Nature's Form of Design
(an excerpt from Van Alstyne and Logan 2006)

The question of control versus influence is the crux of the contrast between human design and natural emergence. Nature does not control; she merely accepts whatever is the best fit. Natural selection, the force that selects, is the result of the aggregate of environmental factors and the attrition of individuals incapable of mating or propagating their organization.

Perhaps human designers can learn from nature new ways to design more effectively. What is her secret? Well to start with she spawned these creatures, life forms that could organize themselves, act in their own self-interest, adapt to changing conditions while continually and relentlessly searching for improvements in the Adjacent Possible, thereby creating new species, new genres and even new taxas. Nature did not actively spawn these creatures—she merely created a set of physical laws, including organic chemistry, which allowed them to emerge though self-organization. And why were these creatures able to achieve this magnificent accomplishment? The answer is so simple it is often overlooked. They had *purpose*—the purpose to propagate their organization. Those that were able to realize that purpose survived, lived and bred, and those that were not able fell by the wayside and were heard from no more.

So what is the bottom line for the designer? Purpose must be the starting point, the motivating factor. Next the materials must be in place, the elements that will go into the design. Then the designer must catalyze the process so that elements of the design self-organize into a pattern that can achieve the purpose or *telos* of the design.

These four elements represent the four causes of Aristotle: material, formal (the pattern), efficient (the designer) and final or *telos* cause (the purpose). The designer is the efficient cause trying to make the final cause—the purpose. Designing is causing.

Econosphere

We remind the reader that the econosphere consists of the symbolic patterns for the organization of materials, energy and human activity that result in the economic units or systems of exchange consisting of businesses, NGOs and governmental agencies.

These patterns of organization represent **information**. The actual physical instantiation of these economic units that are impacted by downward causation from the econosphere are open to energy, matter and information whereas the symbolic patterns of organization that constitute the econosphere are **open** only to information.

The units of the econosphere, which are subsets of culture, behave like organisms, obligate symbionts that enhance the ability of their hosts to secure free energy from the environment to hence contribute to their host's ability to sustain themselves and propagate their organization. The units of the econosphere **evolve** through the Darwinian mechanism of descent, modification and selection, as was pointed out by Johnson and Earle (1987) in their book, *The Evolution of Human Societies*. They have identified a universal pattern in the evolution of political-economies based on society's need for sustenance.

> Subsistence intensification, political integration and social stratification are three interlocked processes observed again and again in historically unrelated cases. Foragers diversify and gradually adopt agriculture; villages form and integrate into regional polities; leaders come to dominate and transform social relationships.... We see the evolutionary process as an upward spiral. At the lowest level the pressure of an increased population on resources evokes a set of economic and social responses that interact to create a higher level of economic effort capable of sustaining an increased population. The process repeats itself until eventually a growing population becomes possible only with the increasing involvement of leadership, with its concomitants of increasing dependence and political development. (Johnson & Earle 1987, 4, 15)

It was through cultural transmission that the features of the previous political-economic system are incorporated into the new political order that arises to deal with the pressure of an increasing population. This is the sense in which the elements of the econosphere evolve through the mechanism of descent. While the older forms descend into the new emerging system there still must be a modification of the old form of social organization to deal with the new challenges. The modifications that arise in the econosphere parallel the punctuated equilibrium of the biosphere. The very first original human economic units were the biological family where the only exchange was between family members. As populations grew and the competition for declining resources grew, new forms of organization emerged. The modification of economic units was initiated by a small cadre of leaders and social innovators that were then selected by the community as a whole. The selection process worked such that those communities that adopted forms of organization that were most fit survive and competed favorably with those communities that adopt forms of organization or maintained forms of organization that were less fit. This mechanism of selection parallels biological evolution. The mechanism of modification differs in the econosphere because it is not random as is the case in the biosphere.

There is another analogy between the evolution of economic-political systems and biological organisms, which is that as new structures emerge they are retained as the complexity of the systems increases. With biotic systems once the heart emerged or once the brain emerged all higher order animals retained these features. Once the spinal column emerged in the simplest and earliest vertebrates all higher order species such as fish, amphibians, reptiles, birds and mammals retained this structure. In every case this structure encased the spinal cord but played a different role in terms of each of the animals' skeletal structure. In the same way the family retained its basic structure despite the many changes of society including changes of the overall political and economic structures of society.

Economic and governmental organizations have a life of their own **agency** by virtue of the institutional will of the organization, which often supercedes the will of individuals within the organization. Citizens do not wish to pay taxes but they are obliged to do so by their government. There is a tradition in modern political theory dating back to Hobbes that the state should be regarded as a person. This way of looking at the body politic incorporates the notion that the state has agency like an organism in the biosphere.

The same applies to other units of the econosphere. A corporation is regarded in law as a person and hence as an agent. In a company the employees cannot pursue their own self-interest but must act in the interest of their firm, which is to be profitable. For NGOs the objectives of the organizations incorporates their agency.

Another from of agency is that a political economy shapes the way in which humans interact with their technology and with each other. Consider the capitalist system during the height of the industrial revolution in which private individuals invested in the means of production and in order to recoup their investment they had to have their machines in operation as much as possible. This led to shift work and the demand on the workers to be punctual, obedient and perform rote work. The agency of the political economic system and the technology of mass production had an agency of its own which forced the factory owners to impose on their workers a non-humanistic work routine.

Economic and governmental organizations are **symbiotic** in that they interact and trade with each other to form political economies and international trading partnerships. It is also the case that businesses and NGOs depend on government agencies in order to function. Collaboration is now seen as not just the cooperation of individuals but also of corporations and other forms of business even to the extent that it is advocated that organization that compete in the market place can still develop collaborative or symbiotic relationships (Logan 2004c).

The analysis that we have just made is reviewed with the following matrix where the M_{ij} are described below:

Matrix of Spheres of Influence and their Properties

C1\R1	Info*	Evolution	Agency	Open*	Symbiosis
1. Biosphere	M11	M12	M13	M14	M15
2. Symbolosphere	M21	M22	M23	M24	M25
3. Technosphere	M31	M32	M33	M34	M35
4. Econosphere	M41	M42	M43	M44	M45

Info* = Information content; Open* = Openness to information and in the case of the biosphere openness also to energy and matter.

M11: The information content of the biosphere is in the form of DNA, RNA, proteins and the other forms of biotic or instructional information as defined in POE.

M12: The evolution of living things as described by Darwin in terms of descent, modification and natural selection.

M13: Living organisms are autonomous agents that act in their own self-interest.

M14: All living organisms are open to matter, energy and information.

M15: All living organisms live in symbiotic relationships with other organisms.

M21: Language and culture are symbolic systems, which carry information and are at the same time a medium for communicating information.

M22: Language and culture evolve by descent, modification and selection through the mechanisms of memes.

M23: Language and culture act as organisms with their own agency as described by Christiansen (1994) for language and Logan (2007) for culture.

M24: Languages and cultures are open to information as this is the mechanism by which they are modified.

M25: Languages and cultures are not isolates but live in interaction with other languages and cultures.

M31: Tools and media that belong to the technosphere are not the physical instantiation of these technologies but the symbolic concepts of their design, i.e. patterns for the construction of the tools or media. As such they are forms of information.

M32: Technologies evolve through the mechanism of descent, modification and selection as pointed out by Basalla (1988).

M33: Technologies have agency by virtue of their unintended effects as pointed out by McLuhan (1964).

M34: Technologies are open to information as this is the mechanism by which they are modified.

M35: Technologies are symbiotic is that the success of one technology depends on the existence of other technologies they co-exist with.

M41: Economic units and governmental agencies are symbolic patterns for the organization of materials and humans and as such they represent information.

M42: Economic units evolve through the mechanism of descent, modification and selection as pointed out by Johnson and Earle (1987).

M43: Economic and governmental organizations have agency by virtue of the institutional will of the organization, which most often supercedes the will of individuals within the organization.

M44: Economic and governmental organizations are open to matter, energy and information but the constituents of the econosphere, which are the symbolic patterns of exchange, are open only to information.

M45: Economic and governmental units organizations are symbiotic in that they interact and trade with each other to form political economies and international trading partners.

Conclusion

Our matrix summarizes the results of our analysis and comparison of the four spheres of propagating organization that impact the human condition. A number of conclusions may be drawn from this data. The first is that the analogy we have drawn between living organisms and language, culture, technology and political-economic systems in earlier chapters is rather robust. We also see that these systems of propagating organization that enter into symbiotic relationships with each other and with us humans are more than extensions of who we are as McLuhan once claimed (1964). These systems are as much a part of us as are any of our organs. Without them we would not be fully human and if they had not evolved and adopted themselves to our changing conditions we might not have survived and certainly not in the numbers that now populate the planet.

One thing our analysis reveals is that information is not a static thing but rather a dynamic process. Although information is a noun it certainly behaves like a verb. Having developed a general overall framework for information we turn in the next three chapters to three domains where information plays a central role. In Chapter 8 we look at the role of information in science and its reliability. In Chapter 9 we look into the future of the book, the medium that has dominated the storage and transmission of information for the past 5000 years. Finally in Chapter 10 we demonstrate how the information processing capability of conceptualization has led to different forms of artistic expression. While these three case studies are not central to our mission of answering the question what is information, they enrich our portrait of the nature of information and the role it plays in human affairs.

Chapter Eight

Science as a Language, the Non-Probativity Theorem and the Complementarity of Complexity and Predictability

In the last chapter when we studied the nature of the impact of the four spheres on the human condition we omitted science from our discussion. In this chapter we will study the nature of science as one of our objectives in this project is to create a dialogue between science and the humanities and the social sciences using information as a bridge connecting them.

In particular we will try to answer the questions:

- What is science?
- How does it differ from mathematics?
- What is the relationship of information to science?
- What is the reliability and truth-value of the information generated by science?
- Can a scientific analysis prove anything?

At the Humanity and the Cosmos Symposium held at Brock University in St. Catharines Canada in January 2000 (see the acknowledgement at the end of this chapter) a number of the participants made statements to the effect that science could prove this or that. During the course of our discussions it suddenly occurred to me that science cannot prove anything but only offer up hypotheses to be explored empirically. This chapter is an elaboration of that thought.

A linguistic analysis and a formal mathematical proof will be presented to show that science cannot prove the truth of a proposition but can only formulate hypotheses that continually require empirical verification for every new domain of observation that is encountered. A number of historical examples of how science has had to modify theories and/or approaches that were thought to be absolutely true and unshakable are presented including the shift in which linear dynamics is now considered anomalous and non-linear dynamics the norm. Complexity and predictability are shown to have a complementarity like that of position and momentum in the Heisenberg uncertainty principle. The relationship of complexity and predictability is also shown to be similar to that of completeness and logical consistency within the framework of Gödel's Theorem.

Science as a Language

Because science is a form of organized knowledge in order to understand the relationship of information to science we need to understand the relationship of information and knowledge. In Chapter 2 we identified information as structured data and knowledge as the ability to use information strategically to achieve ones objectives. The objective of science is to describe nature as accurately and simply as possible. As Einstein opined a theory should be as simple as possible but not too simple. In Chapter 3 we also argued that science and mathematics are languages and therefore part of culture and hence the symbolosphere. Within the framework of this model of the evolution of language, mathematics and science are seen to be distinct languages each with their own unique informatic objectives.

Mathematics strives to solve equations and to prove the equivalence of sets of propositions involving the semantical elements of its language, namely, abstract numbers (such as integers, other rational numbers, irrational numbers, and imaginary numbers), geometrical objects (such as points, lines, planes, triangles, pyramids, vectors, and tensors), sets, operators, etc. A theorem or a proof is a unique syntactical element of the language of mathematics, which we will show cannot be an element of the language of science. A theorem or a proof establishes, using logic, the equivalence of one set of statements with another set of statements, a proposition whose truth is to be established

Atacama Large Millimeter Array

by the theorem. The first set of statements includes axioms, whose truths are assumed to be self-evident and, at times, other theorems, which have already been proven based on the same set of axioms.

Science, on the other hand, establishes the veracity of a proposition using the technique of the scientific method of observation, generalization, hypothesis formulation, and empirical verification of the predictions that emerge from the hypothesis. The scientific method is a unique syntactical element of the language of science. In addition to trying to provide an accurate description of nature science also attempts to describe nature in a systematic manner using the minimum number of elements possible. The description of one phenomenon in terms of another is often claimed to be an explanation. This is one way to interpret this reduction of the number of basic elements needed to describe nature, which is a basic goal of science. Science also endeavors to make predictions that can be tested to establish the accuracy of its models. No matter how refined these processes become and no matter how many reductions and simplifications are made there always remain some irreducible elements that resist explanation or description in terms of simpler phenomena. The process of reduction has to end somewhere. The basic elements in terms of which other phenomena can be described can be thought of as the basic atoms or elements of scientific description (MacArthur, 2000).

Scientists often make use of mathematical language to construct their models of nature, especially in the physical sciences. They often employ mathematical proofs to establish the equivalence of mathematical statements within the context of their models. This has led to the popular belief that science can actually prove things about nature. This is a misconception, however. No scientific hypothesis can be proven; it can only be tested and shown to be valid for the conditions under which it was tested. Each proposition must be continually verified for each new domain of observation that is encountered.

The purpose of this chapter is to make use of mathematical reasoning to show and actually prove that science can never prove the truth of any of its propositions or hypotheses. Our proof is based on an axiom proposed by Karl Popper (1959), namely that a hypothesis, proposition or theory is scientific only if it is falsifiable, namely it has the possibility of being shown to be false by an observation or a physical experiment, in other words it is testable with the possibility of a negative outcome.

For the purposes of our study we need to clarify what we mean when we use the word truth by distinguishing two types of truth, empirical or verifiable truth and necessary or analytic truth. Empirical truth arises from the matching of a measurement with a model and is always approximate to some degree or other depending on the precision of the measurement and the accuracy of the model. Necessary truth arises out of mathematical reasoning or the use of logic and is exact. Although necessary truth is exact its validity depends totally on the basic axioms from which one starts and which one assumes are self-evidently true. At some point one must rely on belief to establish that an axiom is self-evidently true. The necessary truth of mathematics or logic is therefore artificial. The most one can say about the truth of mathematics and logic is that subject to the limitations of Gödel's Theorem it can only demonstrate the equivalence of one set of propositions with another. Mathematics and logic are therefore our very first examples of virtual reality. Empirical truth while less precise than necessary truth at least attempts to describe reality. The scientific models are artificial and are only representations of reality but they do have to measure up.

To establish our theorem, the Science Non-Probativity Theorem, we will make use of Popper's basic axiom, namely, that for a statement or an assertion to be considered as a scientific statement it must be tested and testable and, hence, it must be falsifiable. If a proposition must be falsifiable or refutable to be considered by science then one can never prove it is true for if one did then the proposition would no longer be falsifiable, having been proven true (in the sense of necessary truth), and, hence, could no longer be considered within the domain of science. We have therefore proven that science cannot prove the truth of anything. Any proof of the truthfulness of a proposition would put that proposition outside the realm of science and place it within the domain of mathematics or logic. And as was pointed out by Stephen Clark (2000), "Not all proofs are ever intended as 'necessitations'. So what counts as 'proof' will vary between disciplines and practices."

The Science Non-Probativity Theorem

Let us repeat the above argument as a formal theorem making use of two axioms.

Axiom 1: A proposition must be falsifiable to be a scientific proposition or part of a scientific theory.

Axiom 2: A proposition cannot be proven necessarily true and be falsifiable at the same time. [Once proven true, a proposition cannot be falsified and, hence, is not falsifiable.]

Theorem: A proposition cannot be proven to be true by use of science or the scientific method.

Proof: If a proposition were to be proven to be true by the methods of science it would no longer be falsifiable. If it is no longer falsifiable because it has been proven true it cannot be considered as a scientific proposition and hence could not have been proven true by science. Q.E.D.

Spiral galaxy

In the spirit of the Science Non-Probativity Theorem and our distinction between necessary and empirical truth, we cannot be certain that this line of reasoning is absolutely valid or "true". After all we have just used the theorem, a syntactical element of the language of mathematics to establish a proposition about the language of science. The validity of our conclusion is no greater than that of our starting axioms. Our theorem is not scientifically valid but as a result of mathematical reasoning we have created a useful probe; one that can lead to some interesting reflections and insights into the nature and limitation of science. If it helps scientists and especially the public, who tend to accept the authority of science more or less uncritically, to adopt a more humble and modest understanding of science, it will have served its purpose. The purpose of this exercise was not, as some have suggested, to challenge the usefulness of science or the validity of its methodologies but to clarify the nature of scientific truth and contrast it with the necessary truth of logic.

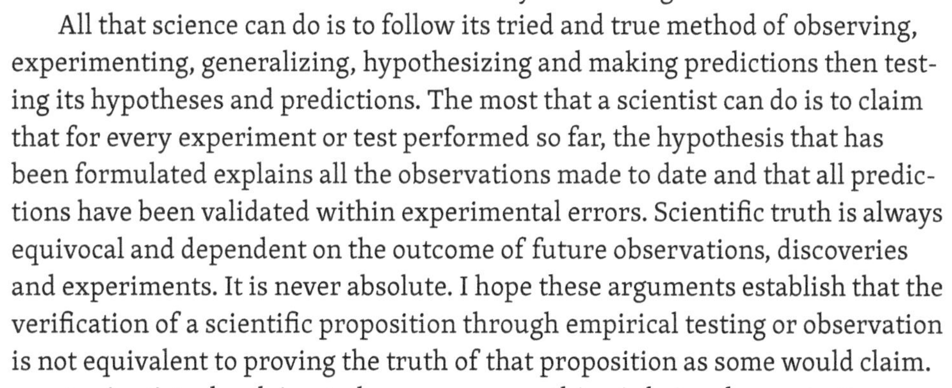

All that science can do is to follow its tried and true method of observing, experimenting, generalizing, hypothesizing and making predictions then testing its hypotheses and predictions. The most that a scientist can do is to claim that for every experiment or test performed so far, the hypothesis that has been formulated explains all the observations made to date and that all predictions have been validated within experimental errors. Scientific truth is always equivocal and dependent on the outcome of future observations, discoveries and experiments. It is never absolute. I hope these arguments establish that the verification of a scientific proposition through empirical testing or observation is not equivalent to proving the truth of that proposition as some would claim.

A scientist who claims to have proven anything is being dogmatic. Every human being, even a scientist, has a right to their beliefs and dogmas in their basic axioms upon which their proofs are based. But it does not behoove a person who claims to be a rational scientist and who claims that science is objective and universal to be so absolute in their beliefs and in the value of their belief system, science. Scientists are not immune to dogmatic and intolerant views as Dr. George Coyne (2000) has pointed out in his talk at the Humanity and the Cosmos Symposium at Brock University, "When the Sacred Cows of Science and Religion Meet".

I believe, it is altogether fitting and appropriate, that scientists should display greater humility and tolerance in the practice of their vocation and calling (Bertschinger, 2000) in view of the lessons to be learned from the following historical vignettes where well established scientific theories and dogmas had to give way to newer ones.

Newton's theory of motion gave way to Einstein's theory of relativity once one considers velocities that approach the speed of light. The Newtonian picture also underwent major revisions with the introduction of quantum mechanics needed to describe atomic systems. Neither the contribution of Newton to science nor the validity of his model of dynamics for non-relativistic and non-quantum events were in any way diminished by these 20th century discoveries. In fact, many elements of Newton's theory survived in both relativity and quantum mechanics and one cannot imagine how these theories could have been formulated without the pioneering work of Newton. Even today's current version of quantum mechanics requires the use of the classical Newtonian Hamiltonian to formulate the energy operator.

Albert Einstein, 1921

Einstein helped to launch quantum mechanics with his explanation of the photoelectric effect in 1905. Despite this pioneering work he turned on the child of his own creation, quantum mechanics, claiming that it is an incomplete theory. Einstein's objections have given way to the acceptance by the main stream of the physics community of probability as being an intrinsic part of our observation of nature due to the Heisenberg uncertainty principle. Einstein's hypothesis that quantum mechanics is an incomplete science can never be disputed or disproved according the Non-Probativity Theorem formulated above. The usefulness of his hypothesis, however, dwindles in the absence of any concrete progress towards a complete non-probabilistic theory of quantum mechanics and atomic systems. And this despite the valiant efforts of David Bohm, Roger Penrose, and others to find the hidden variables or structures that they claim would make quantum mechanics a complete theory. One cannot but help to conjecture that perhaps the reason that these variables are so well hidden is that they do not exist. But this is only my conjecture and belief and not anything that I could prove.

Einstein, Time magazine's man of the 20th century and whose name is synonymous with genius had no problem rejecting one of the elements of his theory of general relativity. He introduced a cosmological constant into his theory in 1914 to describe what he thought at the time was a steady state universe. When Hubbell showed in 1929 that we lived in an expanding universe Einstein immediately dropped this element of his theory. Some contemporary cosmologists have since resurrected it because they find it might serve a useful purpose in their attempts to explain or describe certain specific observations of the cosmos.

Another interesting shift in attitudes within the physics community is illustrated by the recent emergence (pun intended) of chaos theory, complexity, simplicity, plectics, emergence, and self-organizing criticality all of which concern themselves with non-linear dynamic systems. It was once claimed, not very long ago, that the complications that non-linear equations presented were mere details not worthy of attention since the basic equations of motion, while not soluble in closed form, are at least amenable to numerical analysis if one needed to solve these equations. In fact Poincaré showed that there is no unique solution to the 3-body problem.

> When simple laws govern systems with a large number of variables, the underlying order is obscured by our inability to track every component, and it becomes inaccessible to our limited brainpower. Within the last decade this view of the origin of complexity has been strongly challenged.... At the frontiers of today's mathematics are startling paradoxes about the way the world can change. In particular, we now know that rigid, pre-determined, simple laws can lead to behavior so irregular that it is to all intents and purposes random (Cohen and Stewart 1994, 20).

With the availability of computers, especially microcomputers because they provided researchers with low cost computing power that allowed them to play, scientists were able to explore and examine the complexity of non-linear dynamical systems and their sensitivity to initial conditions. As a consequence many interesting results were arrived at and it is now widely recognized that non-linear physics is not a special case or the anomaly of nature but rather the norm that requires detailed attention. The shoe is now on the other foot and it is realized that it is the dynamical systems that can be described by linear equations that are the anomalies or unusual cases. It was only because they could be described in simple closed mathematical equations that they received as much attention as they did.

In light of the Non-Probativity Theorem it is clear that the role of science is to probe and not to prove. It is interesting that the two English words, prove and probe, both derive from the same Latin root, proba, which means prove. The words probability, problem and probable all have the same root. This makes Einstein's rejection of probability in quantum mechanics all the more ironic.

Science, the Language of Metaphor

Science involves the process of representing empirical observations in terms of models many of which are mathematical. These models whether or not they are mathematical are metaphors for and abstractions from nature. The spirit in which scientific models are described as metaphors is the same as that of the proposition that all the words of a spoken language are metaphors. The idea that all communication is based on metaphor is an idea that "has ancient origins in oral cultures and has been repeated and debated through history" (Gozzi, 2000) by Plato, Vico, Keats, Shelley and many modern linguists. McLuhan (1964) quotes Quintillian "Nearly everything we say is metaphor."

Once a scientific model is formulated in terms of some basic axiomatic metaphors, mathematical and/or logical relationships between these metaphors are explored leading to predictions in the form of new metaphors. The relationships between the axioms and the predicted metaphors have the rigor of a mathematical proof but the validity of the model is determined by how well the predicted metaphors match the observations of nature. The most one can say, ala Hume, is that the newly predicted metaphors transformed by mathematics from the original axiomatic metaphors of the starting model make a good match to the observed phenomena of nature. This empirical agreement supports the scientist's model but does not prove that the model is correct because one must leave open the possibility that the model can be falsified or refuted or perhaps just improved.

If as noted above all words are metaphors and all scientific models are also metaphors there is no need to prove that scientific statements are true. One cannot prove a metaphor is true one can only test whether or not it provides a useful description of nature, which leads to greater insights and in the case of science to more predictions or in the case of the arts to deeper insights. It is the natural process of a language to evolve, the same is true of the meaning of words and metaphors. Words are continually bifurcating keeping their old meaning and taking on new meanings. The new meanings, however, carry with them vestigially some of the structure or meaning of their ancestors just as animals and plants vestigially retain structures from their ancestors. Scientific theories, which are made up of metaphors, also evolve and bifurcate into new models, which vestigially retain remnants of earlier theories. Relativity and quantum physics still retain much of classical Newtonian physics. Plus ca change plus c'est la meme chose.

All models are abstractions from nature and hence represent a reduced reality. Mathematical transformations of the abstractions or metaphors of a model may further degrade their accuracy and reduce their match with empirical reality.

The role of science is not to prove or even to explain the phenomena of nature but rather to uncover patterns that relate one set of phenomena to another. The mathematicizing of scientific models and metaphors and the process of subjecting them to mathematical operations has proven to be a successful technique in uncovering these patterns especially when predictions are made that can be observed or measured.

The Complementarity of Complexity and Predictability

The assumption that the metaphors contained in mathematical models used to describe nature can then be operated upon using linear mathematical operators to obtain new relationships among the elements of the model which will then correspond to what is observed in nature is premised on the notion that the relationship between the elements of the model and the elements of reality are linear. This is an assumption or basic presupposition, which cannot be proven mathematically but must be tested empirically and cannot be presumed to be necessarily true.

The effect of a non-linearity between the model and reality can become magnified if the mathematical equations relating the elements of the model are themselves non-linear. A small difference or non-linearity between the mathematical model and the reality being modeled can lead to vastly different outcomes ala the butterfly effect of Lorenz.

Quantum mechanics and the uncertainty principle have taught us that the process of measuring nature at the atomic scale changes the phenomena we are observing and scrutinizing. Something similar happens with complex processes, which

> generate counterintuitive, seemingly acausal behavior that's full of surprises.... Complexity is an inherently subjective concept; what's complex depends upon how you look.... Whatever complexity such systems have is a joint property of the system and its interactions with another system, most often an observer and/or controller."
> (Casti, 269–71)

The modeling of nature using metaphors introduces a new level of uncertainty in matching one's model with nature especially when one attempts to represent the non-linear phenomena using classical pre-chaotic physics. Paradoxically the introduction of chaos has led to the discovery of new patterns and insights into the nature of non-linear dynamic systems ranging from the behavior of ecosystems to the origin of the universe.

Within the new physics of chaos or complexity theory, the chaos or the uncertainty associated with not being able to make predictions of the behavior of non-linear systems leads, as Prigogene first suggested, to new levels of order. The Heisenberg uncertainty principle in quantum mechanics, which does not allow the simultaneous determination of the position and the momentum, leads to an understanding of the wave nature of particles and the particle behavior of light and by association to an understanding of the wave behavior of the probability amplitudes needed to describe atomic and sub-atomic particles and make predictions about their behavior. Just as momentum and position (or energy and time) play complementary roles in the Heisenberg uncertainty principle, complexity and predictability seem to play a similar complementary role. Complexity and predictability are hard to quantify in this context unlike the uncertainty in momentum and position, Δp and Δx, in quantum mechanics. But it is the case that one cannot at the same time take into account all of the variability of a non-linear system and still formulate the equations that will predict the behavior of the system.

Isaac Newton, Godfrey Kneller, 1689

> Indeed, any theory of complexity must necessarily appear insufficient. The variability precludes the possibility that all detailed observations can be condensed into a small number of mathematical equations, similar to the fundamental laws of physics.... If, following traditional scientific methods, we concentrate on an accurate description of the details, we lose perspective.... Chaos theory tells us that many simple mechanical systems, for example pendulums that are pushed periodically, may show unpredictable behavior. We don't know exactly where the pendulum will be after a long time, no matter how well we know the equation for its motion and its initial state (Bak, 9–11).

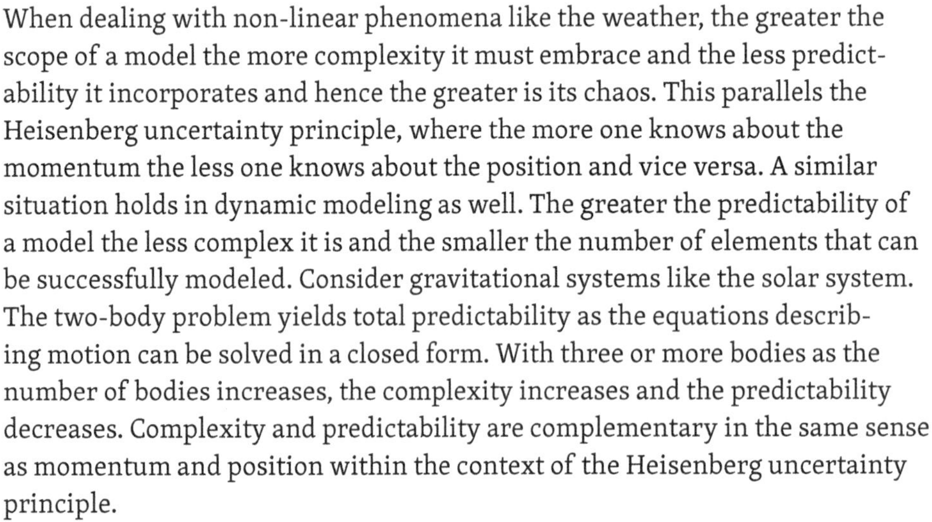

When dealing with non-linear phenomena like the weather, the greater the scope of a model the more complexity it must embrace and the less predictability it incorporates and hence the greater is its chaos. This parallels the Heisenberg uncertainty principle, where the more one knows about the momentum the less one knows about the position and vice versa. A similar situation holds in dynamic modeling as well. The greater the predictability of a model the less complex it is and the smaller the number of elements that can be successfully modeled. Consider gravitational systems like the solar system. The two-body problem yields total predictability as the equations describing motion can be solved in a closed form. With three or more bodies as the number of bodies increases, the complexity increases and the predictability decreases. Complexity and predictability are complementary in the same sense as momentum and position within the context of the Heisenberg uncertainty principle.

The decrease in the predictability of a model of a non-linear dynamics system because of the increase in chaos does not represent a shortcoming of the model but rather an attempt to be complete by including the full complexity of the phenomenon being represented. In the spirit of the Non-Probativity Theorem there is no reason to believe apriori that a model representing nature should be both complete and totally predictable. Gödel's Theorem can serve as a possible model to better understand the complementarity of complexity and predictability. Gödel's Theorem states that a mathematical system cannot be both complete and logically consistent at the same time. If we think of predictability of phenomena as a form of logical consistency with the basic laws of nature and consider complexity as a form of completeness then Gödel's Theorem also supports the notion that total complexity or completeness of a model precludes complete predictability.

The rejection of chaotics and complexity theory by adherents of the older paradigms of Newtonian physics, relativity and quantum mechanics is due to the fact that the new physics places limitations on the predictability of nature. Einstein critiqued quantum mechanics when he proclaimed, "God does not play dice". The new physics is even more disturbing to this new generation of skeptics who have to contend with the notion that not only does God play dice at the atomic and sub-atomic level but he also plays it at the macro level. Even though the interactions of complex classical systems are causal they are not predictable because of their complexity and their non-linear dynamics and therefore seem random like quantum effects. As a consequence of this one must give up on the notion of the prediction of certain phenomena at the macro level, something that not even quantum mechanics required despite the fact that it made use of probability at the micro level. Equally disturbing to some is the fact that the very existence of human life might also be the result of a random role of the dice.

The new physics places limitations on the ultimate ability of science to predict certain phenomena critical to human survival such as the weather and large scale climatic change no matter how sophisticated our computational skills become. Buying into the new physics requires accepting the fact that some problems are intractable. This requires a new level of humility on the part of science, which has enjoyed a period of unprecedented success for over 500 years in which it has been able to describe and explain almost every phenomenon it has encountered. Are we willing to sacrifice the sacred cow of predictability and accept a more modest role for ourselves in our quest for understanding our universe? Will we accept a worldview in which chaos and non-predictability is regarded as natural outcomes of the complexity and diversity of our universe, a richness, which gives rise to this dilemma. I believe that the next generation of physicists will happily sacrifice this sacred cow and move on to a higher and deeper understanding of nature in much the same way that the Hebrews gave up the golden calf at Sinai and embraced ethical monotheism, but not without becoming stiff-necked, however. The only solace that can be offered to those who are disturbed by the lack of predictability of the new physics is that events are still causally connected but that at the edge of chaos where self-organizing criticality takes place science will not be able to determine which new form of equilibrium will emerge.

Conclusion

In this chapter we have attempted to show the strengths and limitations of science when regarded as a language with its dual role of communication (description) and information processing (predictability). The Non-Probativity Theorem underscores a long held belief that scientific truth is not absolute but always subject to further testing. We have tried to link the limitations on predictability within the framework of the new physics of non-linear dynamics with the Heisenberg Uncertainty Principle and Gödel's Theorem. We have suggested that the chaos and non-predictability of complexity theory allows a more complete and fuller description of nature.

Acknowledgments: I wish to thank the organizers and participants of the Humanity and the Cosmos Symposium, January 20–22, 2000 sponsored by Brock University and the Brock Philosophical Society where the ideas for this chapter were incubated. I would also like to thank George Coyne personally for his powerful keynote address, which inspired the ethical dimensions of this chapter. For their insightful presentations and lively discussions I would like to individually thank, Hugo Fjelstad Alroe, David Atkinson, Richard Berg, Edmund Bertschinger, Leah Bradshaw, Bruce Buchanan, David Crocock, G.E. Dann, Darren Domski, George Ellis, David Goicoechea, Anoop Gupta, Calvin Hayes, Daniel MacArthur, K. McKay, M.J. Sinding. I would also like to acknowledge my colleagues on the Media Ecology listserv for a recent stimulating discussion of metaphors with Jim Curtis, Raymond Gozzi, Randolph Lummp, John Maguire, Eric McLuhan, Lori Ramos and Lance Strate. I also wish to acknowledge the help of my former colleague in the Physics Department at the University of Toronto, Prof. Ken McNeill of blessed memory.

SUNTORY
Fe
出入
住友中之島

Chapter Nine

What is a Book? Past, Present and Future From the Clay Tablet to the SmartBook

The book represents a technology that has for 5,000 years been the principle medium for the storage and transmission of information. It is only with the emergence of digital information with computing and the Internet that the printed book has had any competition as the chief medium for the preservation and dissemination of information.

> *"Every exit is an entrance to somewhere else."*
> *—Tom Stoppard*

Although the focus of this book has been to answer the question "what is information?", I believe it is appropriate to address within this volume the questions: What is a book?; What competition digital forms of information will present?; and How will digital information reshape the book? We will address these questions making use of a media ecology perspective as developed by Marshall McLuhan.

But before peering into the future of the book let us first take a look in the rear view mirror and outline a short history of the book. Generally when one uses the term "book" one immediately thinks of the printed codex book in which a number of pages containing text printed in ink on paper are bound together. We will use a more general definition of the book as a collection of written text. With this definition we may claim that the first books emerged in Sumer where writing was invented some 5,000 years ago as clay tablets. These tablets were produced using a cuneiform code that was pressed with a stylus into wet clay and was then subsequently baked to create a permanent record. A number of these tablets have survived to this day. One might also include in the general category of books texts that were carved into stone as was the case in the legal stelae of Babylon such as the Hammurabi code and the hieroglyphics or "sacred writings" on the walls of ancient Egyptian monuments. These are not books in the usual sense of are understanding of this medium because they lack portability and were only meant for public reading.

The next format in the evolution of the book was ink written on long sheets of parchment, papyrus or paper that was then rolled into a scroll. The first scrolls were probably produced in ancient Egypt shortly after the emergence of writing in Sumer. The use of scrolls spread throughout the world and were the principle medium for books until the invention of the codex format by the Romans in the first century of the Common Era. The production of new scrolls survived in Europe into the medieval period when they were finally replaced by the codex format. The Hebrew Torah is an example of the scroll, which is used and read aloud to this day in the religious service of synagogues throughout the world.

The next step in the development of the book and something that more closely resembles today's printed book was the codex format of pages of text written with ink on parchment, papyrus or paper. Until the invention of the printing press all books were hand written. The first printing presses emerged in China in which text was carved in reverse on wooden blocks. The very first printing press in Europe was developed by Laurens Janszoon, surnamed Coster in Haarlem in the Netherlands around 1420 making use of the Chinese technique of using carved wooden blocks to which were added a few fonts of single letters. The first movable type printing press was the invention of Johannes Gutenberg of Mainz Germany around 1440. This invention created an explosion in the number of books manufactured and a revolution in culture leading to the Reformation, vernacular literature, nationalism, the scientific revolution (McLuhan 1962, Logan 2004a). The development of the press was followed shortly thereafter by the innovations of Aldus Manutius who invented the italic font that allowed more text on a page and who began publishing books in a much smaller and less expensive format that increased both the portability of the book and its popularity. Other advances in the technology of printing

Gutenberg Bible

that led to still cheaper books and the emergence of public education during the industrial revolution greatly increased the size of the reading public. The introduction of the paperback book in recent times was another step that increased the size of the reading public.

Printing press, 1568

With computing, word processing, and desktop publishing the production of books was once again given a boost. The use of computers to store texts plus the advances in postscript printing and photocopying led to the print-on-demand book. These developments did not change the format of the book, which still retained its codex form of ink printed on paper.

The most dramatic transformation of the book and the one that no longer required the codex format of ink on paper was the emergence of the ebook in which the content of a book sitting in a digital format on a memory device is delivered to the screen of a reader. The screen could be embedded in a computer, a smart phone or a dedicated ebook reader like Amazon's Kindle. This development has led to the claim by champions and enthusiasts of the ebook that the printed codex book faces imminent obsolescence. We believe that they are greatly mistaken. The book as we will demonstrate shows no sign of giving up the ghost and in fact is thriving as it never has before. So why have some scholars suggested that digital information in the form of personal computers and the Internet threaten to make the codex book of ink on paper obsolete when there is no end in sight for the book? That is a good question.

The reason is that some scholars believe that the Internet, the Web and ebooks will spell the end of the medium of the printed book, a medium that can be traced back to the emergence of writing 5,000 years ago and a medium that has survived the arrival of microfilms, television, personal computers, the Internet and the World Wide Web. The skeptics of the printed book's survival will not dispute that written text will survive but they think that text written with ink on paper sheets that are bound into a codex format, i.e. what we commonly call a book, will give way to purely electronic forms of text delivered over the Internet and the World Wide Web and read on electronically-configured screens. They further claim that the legibility of text on a screen will steadily improve with time so that it will slowly reach the level of ink on paper, a point that we will dispute below. We intend to show that these prophesies of the book's doom are greatly exaggerated and are due to a lack of a deep understanding of how books and electronically-configured screens work at the neurophysiological level. That is not to say the book will remain totally unchanged. No, as we will show the book is destined to take on new forms and achieve new levels of functionality.

I believe one of the reasons for the pessimism of some regarding the future of the book comes from their belief that what happened in the recorded music industry where there has been a sharp decline in the sale of CDs will also happen in the book publishing business. The recorded music industry has been transformed from the producers and distributors of a product in the form of a CD in a jewel box to the providers of a service. Many music stores and chains of music stores have had to close their doors because they could not compete with the likes of iTunes. Because of the downloading of music files recorded music has become a service and not a product. Will the same thing happen to book-length texts?

There is a major difference between the book and music industries. Music is not something that needs to be studied carefully to be fully appreciated. Text, however, requires careful study and close reading and that is why the format of access is extremely important. There is a neurophysiological dimension to reading text on paper versus an electronically configured screen that entails the fact that the two sides of the bicameral brain process information differently. The left-brain is involved in the analytic production and interpretation of both spoken language and written text. The right-brain, on the other hand, is involved in the synthetic processing of spatial and musical perception. Reading a text of ink on paper is strictly a left-brain process. The reading of text on a screen no matter how good the resolution involves both left- and right-brain processes. When a reader is confronted with text on a screen she must first reassemble the mosaic of pixels that represent the letters of the text to form an image of the letters. This is a right-brain process of spatial perception. Readers must then read the text that the right-brain has assembled but using the analytic processes of the left-brain. This is why reading on a screen no matter how good the resolution is a complicated activity. The right-brain converts the pixels into letters and the left-brain converts the letters into words and sentences. There is a lot of traffic through the corpus collosum.

This is why professional writers who produce high quality text will compose their text using word processing, as this does not require using the screen. They will even do some editing on the screen but when it comes time to edit their final draft of the manuscript they will print out their word processed text and make their changes and edits on paper and not on the screen. In June 2009 I surveyed two groups of writers, totally 42 authors in all at the BookCampTO unconference in Toronto and the Media Ecology Association conference in St. Louis and they all agreed that this was their procedure. Editing a book length document requires being able to take in the whole text at once and nothing beats the codex format for being able to see one's book-length-manuscript all at once. Furthermore scholars who have a long text to study that they can access online will still print out the file so they can do a close reading on paper. It is nearly impossible to do a close reading on a screen for anything more that a short email or a single page of text. There was a reason the scroll format for text was replaced by the codex. It is somewhat ironic that we have had to return to scrolling with the ebook.

I am sure the book will survive in its traditional codex form and that, at the same time, new forms will emerge like the ebook, which is quite popular and in new forms such as the SmartBook, which we will describe later at the end of this chapter. There will also be changes in the institutions that support the book such as publishing houses, bookstores and libraries as well as the institutions that are supported by the book such as schools and research institutes. I believe they will undergo changes, which we will briefly describe, as they are the middlemen between authors and their readers and, as we know, the Internet has not been kind to middlemen.

Because I refuse to see an end to the book, some readers might gather the impression that I am a dyed in the wool traditionalist and possibly a Luddite that is opposed to the new digital technologies. Let me put that notion to rest right away by revealing that in fact I believe that in addition to the survival of the ink on paper codex book and the increased popularity of the ebook that new hybrid forms of printed and digital books will emerge. In fact I am working with an international team based in Toronto on developing a hybrid medium, the SmartBook or sBook, consisting of a codex book, a smart tag, the Web, a device that can both read the smart tag and access the Web and a software-based recommender system. The SmartBook to be described below will combine the readability of a codex book, the searchability of an ebook, the networkability of a blog, and the AI capability of a recommender system. So rather that predicting the demise of the book I am suggesting a new era of increased functionality, usability and popularity for the book and the emergence of new forms of the book.

Why Books Survived the Threat of Microforms and Why They Will Survive Digital Media

Microfilms or microforms reduce a whole book to a few small slides that contain the text in miniature that can be enlarged and projected onto a screen with a microfilm reader. They were at one time being touted as a solution to reduce the burgeoning collection size of a library to a manageable dimension. Unfortunately this technology never realized its promise largely because microforms were not a satisfactory medium for reading. Nothing can beat the format of the codex book for the ease of reading as the following passage so eloquently proclaims:

> The book in its traditional form is a memory machine of surprisingly compact and enduring power. It carries in its bindings, its covers and the materials out of which it is made traces of its origin and travels, both as an artifact and as a repository of images and ideas. As a physical object it has what the 20th-century philosopher Walter Benjamin called an "aura," consisting of the host of ritual and metaphysical asso-

> ciations it calls to mind. When some of us recoil from the idea of the digital library, what we mostly fear is the loss of this aura.
> (Matthews Battles, *Boston Globe*, Dec. 26, 2004)

Amazon Kindle 3

I see no conflict between the physical book and its digital incarnation. The book is undoubtedly the best way to read the book especially if one plans to read it from cover to cover. The digital format, however, can be extremely useful, especially if one is using it for reference purposes or if one wishes to pull a quote from the material one has already read in a traditional codex format. I believe that in the not too distant future that scholarly books will be published in a hybrid of both formats.

Who Started the Rumor That the End of the Book in Near?

Where did this notion that the end of the book is near come from and how old is this notion. Some will claim that it was the advocates of the new digital media who first proclaimed that the era of the book was coming to an end. In actuality the first, to my knowledge, to proclaim the obsolescence of the book was Marshall McLuhan (1964) in his book *Understanding Media* published in 1964. What is ironic about this is that McLuhan (1962) who wrote *The Gutenberg Galaxy* was a great champion of the book. In *Understanding Media*, McLuhan documented the radical changes that took place as a result of the transition from the age of literacy and mechanical technologies to the era of electric media. He was alarmed by the trends that he observed particularly the negative effect that television was having on literacy. He sounded a warning: "The electric technology is within the gates, and we are numb, deaf, blind and mute about its encounter with the Gutenberg technology (McLuhan 1964, 32)." Actually he expressed his concerns for the threat of electric technology to literacy eight years before writing Understanding Media (ibid.) and only six years after the introduction of commercial television in North America. He wrote in the revolutionary journal, Explorations, that he co-edited with Ted Carpenter the following:

> It is the almost total coverage of the globe in time and space that has rendered the book an increasingly obsolete form of communication. The slow movement of the eye along lines of type, the slow procession of items organized by the mind to fit into these endless horizontal columns—these procedures can't stand up to the pressures of instantaneous coverage of the earth (McLuhan 1954).

Unfortunately McLuhan did not live long enough to see the reversal of this trend, which began with the emergence of the digital media of personal computers, the Internet, email, text messaging and the World Wide Web, which unlike television, radio and the movies embrace the Gutenberg technology as its content. McLuhan was the first of many scholars to warn of the dangers that television, an anti-intellectual medium, posed to literacy. He was a crusader who likened himself to Pasteur fighting a pestilence: "I am in the position of Louis Pasteur telling doctors that their greatest enemy was quite invisible, and quite unrecognized by them (McLuhan 1964, 32)."

McLuhan was correct to suggest that television posed a threat to literacy during the time he was active as a scholar from 1954 to 1980, the year he passed away. I believe that the threat he identified is not as great today as it was in his time. While TV still poses a threat to literacy and for some children it is having an adverse effect on their ability to read and write digital media has created a new environment, which actually promotes literacy. Since McLuhan first issued his alert over 50 years ago regarding the dangers of television vis-à-vis literacy a new antibiotic has been developed which counteracts the harmful effects of TV and that antibiotic is the emergence of the interactive "new media" of personal computers, the Internet, email, text messaging and the World Wide Web. The content of these media is text, which acts as an antidote to television's harmful effect on literacy.

Children are returning to the Gutenberg galaxy but a digitized Gutenberg galaxy embedded in email, on the Web or in an eBook. The content of the Gutenberg press was the written word assembled by movable type and printed on paper. The content of the digitized Gutenberg galaxy of the Web, blogs, ebooks and email is the word-processed word which still utilizes Gutenberg's mechanism of movable type with recycled and reassembled fonts of the alphabet except in this case the fonts are generated electronically in a wide variety of styles from classical Times Roman to modern sans serif Helvetica instead of being created by pouring hot lead into a mold.

One of the threats to literacy that McLuhan first identified was the speed at which electronically configured information traveled. He spoke of "the electric environment of instant circuitry (ibid., x)" and wrote "electric speed is bringing all social and political function together in a sudden implosion that has heightened human awareness of responsibility to an intense degree (ibid., 20)." He worried that "the action and the reaction occur almost at the same time" so that we lose the detachment with the information we are dealing with and hence "the power to act without reacting (ibid.)." As a consequence, McLuhan believed that we live mythically with electric media. "Myth is contraction or implosion of any power, and the instant speed of electricity confers the mythic dimension on ordinary industrial and social action today (ibid., 38–39)."

I believe that McLuhan's concerns expressed 50 years ago with the speed up of information are no longer a problem because we are now able to slow down the flow of information because of the ease with which we can print out electric information on paper. This gives us the time to reflect on the meaning and significance of a text. McLuhan's thesis was valid at the time he wrote it. He correctly described the information environment that surrounded him. The figure of "the electric speed of information" changes now within the new ground of the widespread access to computing and to cheap and fast printers. In the spirit of McLuhan's emphasis on the figure/ground relationship we must re-evaluate the impact of the electric speed of information. As he pointed out, "No medium has its meaning or existence alone, but only in constant interplay with other media (ibid., 39)."

A radical shift in the effects of electrically configure information has occurred since the time McLuhan first predicted the demise of literacy. With the emergence of personal computers, printers, print on demand, the Net and the Web we have the best of both the Gutenberg galaxy and digital information. Books continue to thrive and we can print out information that comes to us by email or the Web. It is a fact that most people will not read a long text file on the screen of their computer but will print out the text and read it on paper. Reading text on a screen when reviewing emails or surfing the Web is fine but most people prefer reading ink on paper rather than reading on a screen by reassembling pixels when it comes to a long text especially one of book length. This is why I believe that literary culture is returning to pre-television levels and the book is assured of a long lifetime. That is not to say that the form of the book will not change as it has in the past. In the next section we will describe the evolution of the written word and the variety of the forms of the book in the pre-digital era after which we will describe the emergence of new forms of the book with digital information.

The Evolution of Written Material and the Book

We would like to propose that all forms of written material are a book of one sort or another. The concept of the book then incorporates the following early forms: clay tablets, Egyptian writing on walls, Babylonian stelae, hand written scrolls of ink on parchment like the Torah (the first five books of the Bible) used in Jewish religious services, and hand written codex books of ink on papyrus, vellum and later paper. The modern book emerges with the printing press, which at first is a printed version of the manuscript book. With the innovations of Aldus Manutius, the italic font emerges as well as the portable book that can fit into a saddlebag. With power driven printing presses (first steam and then later electricity) the price of books dropped and allowed the mass distribution of books, magazines, journals and newspapers. There is a wide spectrum of text materials that may be considered as examples of books. The boundary between a book by a single author, a book by multiple authors on a single topic, a journal focusing on a single topic, a journal focusing on a discipline or a general subject, a magazine and a newspaper begin to blur. There is no clear boundary between a book and a journal. One can also ask where is the boundary between a book and a pamphlet for example. We will refer to all printed reading material distributed in multiple numbers as a form of "book" while retaining the term book for the canonical codex format of printed material sold in bookstores and collected in libraries. But now we can generalize written material even more and include into this category of "book" digital books in terms of ebooks and actually all forms of text files one can access on the Internet since they belong to the set of objects that are mass distributed written text.

Another argument in favor of the folio book is the ease with which one can go back and reread passages to make sense of what one is reading. This is why the folio book won out over the scroll format of the Torah because it is easier to access information in the folio format compared to the scroll format. Reading a long document on a computer screen is basically using an electronic scroll. Come to think of it we describe reading text on a screen as scrolling.

The Possible Decline in the Use of Books

As we have argued the book is alive and well and will survive the onslaught of digital media. However, we can expect to see a decline in the amount of time spent with traditional books that are ink on paper in the codex format. As McLuhan foresaw, the electrification and then the digitalization of information has resulted in new patterns of information usage. The percentage of time that people spend with books will decline because as new media emerge they invariably crowd out the older media. There is only so much time in a day and time spent with new media will translate into less time for books. The time spent with books might decline but the time spent reading might actually increase because of all the reading involved with "new media" from use of the Web, blogs, email, text messaging, instant messaging and reading the text associated with electronic games. One can expect as a result of these new patterns of usage that books will start to grow shorter in length rather than longer as readers of digital fare become use to smaller hits of information.

Digital Media Actually Encourage Book Writing as well as Distribution and Sales

Although today's traditional book looks identical to the books of the past today's books are actually a hybrid technology in that the information which forms their content has been gathered, written, edited, and typeset using electronic media. Books are easier to write and produce today than they once were because of digital technologies. An index of the health of the book is the sheer number of books that are published each year. In the early 1970s in the United States approximately 40,000 new titles were published each year. That number has more than tripled in recent times indicating that the book is alive and well. I believe that computing and the Internet have done more to encourage the reading and writing of text on paper than it has discouraged. In fact the explosion of academic books and journals have made it difficult for most academic libraries to acquire complete collections of such items. They rely in many instances to electronic versions of journals. As new journals emerge these days they for the most part elect an electronic format and many journals that were formerly printed have also gone strictly electronic. arXiv is an archive for science and mathematics preprints that can be accessed on the Internet. Some scientists choose to publish only in arXiv and do not bother with submitting their papers to a peer reviewed scientific journal. And you can be sure that the scientists that read these pre-prints on arXiv print them out so they can study them carefully.

Another way in which digital technologies support the book is through their distribution by the on-demand printing of books using multifunction printers that are able to store large amounts of digital information. One way this works is that a publisher can print a small run of books initially and once the initial run is sold out the publisher can then print books on demand as a sale is made. Another way that on-demand printing supports the production and distribution of books is that an author no longer needs a publisher but can turn to a service like lulu.com which will take an electronic file from an author and print copies of their book on demand as orders for the book are received.

In addition to the on-demand printing of books there is also the audio book in which a digital recording in made of a reader reading a book out loud. The digital recording can then be played back on a CD player or an MP3 player. This format is used by people who are not sighted or severely dyslexic or by people who want to listen to the book while doing something else like driving a vehicle.

Another boost to the marketing and sale of books are Web sites that either allows books to be purchased online and/or the contents of a book to be examined. Two leaders in this field are Amazon.com and Google Book Search both of which allow the visitor to their site to look inside a book. Amazon allows the visitor to purchase the book online and Google Book Search connects the visitor to an online bookseller including Amazon among others to facilitate the sale of the book. All in all I would say that digital technologies do more to promote the use and production of books than to discourage their use.

The Future of the Book

The book continues to evolve as new means to distribute text are developed. Can an archived listserv discussion be considered a book? What about a blog, is it a kind of book that grows with time? The classical codex book and these contemporary forms of text delivery will converge just as television has converged with the Internet and social networking as news and public affairs programs request comments and questions from the their listeners via email.

All media have increased variety and choice. When literate society made the transformation from hand written manuscripts to printed books there was an explosion of variety and choice as it was so easy to produce a book in multiple copies that could be easily transported. A similar explosion is taking place with the "new media" because of the ease with which media products whether they are text, audio or video can be duplicated and transmitted. As a result material for which there is little demand can still be made available by the operator of a Web site and still turn a profit. This phenomenon has been identified by Chris Anderson (2004), who has given it the name of "the long tail." One can talk of a new class of long tail books that are easily produced and are intended for a small audience as are those published by lulu.com. Dave Gray has published a book, *Marks and Meaning*, as a work in progress using lulu.com asking his readers to help him complete the book. His book is being authored as a wiki. Many books are written by multiple authors who are able to work together because of the ease of digital communication.

The SmartBook

I would like to end this chapter by describing the SmartBook project that Greg Van Alstyne and I are heading up at Strategic Innovation Lab at OCAD Universityand involves an international team of collaborators including Dave Gray, CEO and Founder of XPLANE, the visual thinking company; Peter Jones and James Caldwell, Toronto-based designers; Ramon Sangüesa, head of innovation at CitiLab Barcelona; Carlos Scolari, communications professor in Barcelona, librarian Kathy Kawasaki and U. of Toronto based professors Gale Moore in Sociology and Matt Ratto at the i-School. We are proposing a new format for books by embedding a "smart tag" into a standard printed codex or folio book that points one to a Web site that has the text of the book in a digital format. As a consequence the SmartBook is readable, searchable, networkable and smart. It is a hybrid of the codex book and the ebook.

The SmartBook is readable because of its codex format and the fact that ink on paper is the best way to read text. It is searchable like an ebook because the smart tag points to a digital form of the book's content on a Web site. It is networkable because the Web site or blook containing the digital form of the book can be used for readers to share ideas about the book with each other and the author as is done on a blog. The author can comment on the readers' remarks and also update her book either on the basis of readers' comments or because of new developments in the field of study that the book addresses. Finally, the book is smart because it knows what the reader wants to know and it can recommend what parts of the book are of particular interest to the reader. It achieves these smarts because the device used to access the Web site can incorporate a recommender system that has a profile of the reader's research needs and information interests. The recommender system is therefore able to highlight those portions of the book that will be of particular interest to the reader.

The SmartBook system also allows a codex book to incorporate the advantages of hypertext through the Web interface. One would read the codex book with a Web access device close at hand. The author could indicate places in both the printed and the digital form of the text where one could jump to another part of the book or to another source of information on the Web, which the reader could access through their browser and thereby enrich their reading experience. This means that a book primarily consisting of text could be extended on to a Web site where the reader could access illustrations, photos, as well as video and audio clips.

The SmartBook represents a third option for book publishers in addition to the standard printed codex book (option 1), and to the various digital formats such as an ebook or a book on a CD-ROM (option 2). Options 1 and 2 have their unique advantages. The chief advantage of the printed codex or folio book is that it is the best form yet devised for readability. The codex format is also better suited for quickly browsing the book to get the feel for it especially if the book has a detailed table of contents and index and is written so that the contents of the book are summarized in the first few pages of the book. The advantage of the digital book, on the other hand, is that it is the format of choice for searching and researching. The ebook has the additional advantage that for a complex subject one can easily search the content of the book for topics of particular interest to the reader and thereby tie together related themes. This is particularly true if the ebook is written taking advantage of hypertext.

The SmartBook has all the advantages of both options 1 and 2 and in addition it can customize the content of the book for the specific use of the reader and it can create a forum for the discussion of the book. This feature is particularly useful for books that are written and used for research. There is less of a need for the SmartBook format for a novel or a book of poetry unless that book is a classic that is frequently studied by scholars and students. But even for books that are purely literary the ability of readers to network with each other and possibly the author would be a distinct advantage.

If SmartBooks succeed in penetrating the market they will have an enormous impact of book publishing, booksellers, libraries and schools. Book publisher will not only have to print and distribute codex books they will also have to maintain a Web site for each book they publish. As the number of SmartBooks increases there will be an impact on libraries. Imagine a library of SmartBooks in which a user enters with their recommender system on their notebook or PDA and are directed to those volumes that are of most interest to them. Part of the function of the reference librarian will be taken over by the SmartBook system. The library edition of the SmartBook will have to have a Smart Tag or RFID tag that can transmit a radio signal over a long distance. SmartBooks will also impact bookstores. Imagine walking into a bookstore with one's recommender system and being directed to the books one would want to buy. How convenient!

The School Book

The use of books in schools will not change much at the early grades. I cannot imagine young children learning how to read with an ebook. The need for textbooks in schools is one certain application that will prevent the disappearance of books no matter what developments take place in the world of digital text. Because of the neurophysiological effect of reading from a pixilated screen the codex book is sure to survive in the school system. The use of ebook and the SmartBook in high schools and universities is another matter. One can well imagine that they will find many different applications in these institutions.

Laws of the Media (LOM)

McLuhan developed a set of rules, which he called the Laws of the Media (LOM) (McLuhan 1975 & McLuhan and McLuhan 1988) for studying the effects of media or technologies, which specifically illustrate their counterintuitive nature. A LOM consists of the following four laws:

1. Enhances:	Every medium or technology enhances some human function.
2. Obsolesces:	In doing so, it obsolesces some former medium or technology, which was used to achieve the function earlier.
3. Retrieves:	In achieving its function, the new medium or technology retrieves some older form from the past.
4. Flips into:	When pushed far enough, the new medium or technology reverses or flips into a complementary form.

To gain a deeper insight into the nature of the codex book and the SmartBook we are proposing let us consider the LOM for these two media.

LOM for the Codex Book

1. Enhances:	The storage of and access to information
2. Obsolesces:	Oral tradition and myths
3. Retrieves:	Memory
4. Flips into:	The ebook and/or the SmartBook

LOM for the SmartBook

1. Enhances:	Codex book
2. Obsolesces:	Traditional library
3. Retrieves:	Reference librarian
4. Flips into:	SmartLibrary

Conclusion

Rather than the book coming to an end or being obsolesced I believe the book is about to enter a new chapter in its history that combines the advantages of the printed codex book and the digital or ebook.

Chapter Ten

Non-verbal Information and Artistic Expression in the Symbolosphere and Its Emergence through Secondary Perception

One can roughly classify human communication and forms of information as being either verbal or non-verbal. We have already examined the various forms of verbal information in the symbolosphere as was the case with speech, writing, mathematics, science, computing, the Internet, and the conceptual and symbolic forms of the technosphere and econosphere.

In this chapter we will complete our study of what is information by examining the origin and nature of the non-verbal forms of information and communication inherent in the artistic expression through music, dance and the plastic arts of painting, sculpture, and photography. Some modes of information like film, theatre and opera contain both verbal and non-verbal forms of information. The interesting question we will consider in this chapter is: What is the connection between verbal language and non-verbal artistic expression both of which are symbolic? One hint of a connection is the fact that humans are the only animals that possess verbal language and express themselves artistically through visual images and music. We think this is no coincidence. We plan to demonstrate that the information processing capability of conceptualization played a key role in the emergence of artistic expression.

We will show in this chapter that the origin of artistic expression is linked to the origin of verbal expression through what we term secondary perception, which we define as the perception influenced by verbal language and conceptual symbolic thought. Primary perception, on the other hand, is the perception experienced by non-human animals and our hominids ancestors before the advent of verbal language. As was reviewed in Chapter 3 the Extended Mind model posits that before verbal language hominid thought was purely percept-based and with verbal language human conceptual symbolic thought emerged. In *The Sixth Language* (Logan 2004) it was posited that spoken language and symbolic thought evolved into writing, mathematics, science, computing and the Internet in response to the information overload of the previous languages. One form of symbolic thinking and communication that was not addressed, however, was artistic expression in the visual arts and music, which entails a combination of visual or auditory perception and concept-based thought. I have often been asked by readers of *The Sixth Language* why I had not included the visual arts and music in my theory and weren't these forms of expressions also languages. My answer has always been: yes, they are languages also but they are non-verbal languages and I do not understand the relationship between them and the six verbal languages.

I now believe I have found a link between the verbal and non-verbal languages. In the Extended Mind model the emergence of verbal language is linked to the bifurcation from percept-based thinking to concept-based thinking where the words of verbal language represents our first concepts. The insight that led to this current attempt to link the verbal languages with those of artistic expression was the realization that the visual arts and music are both percept-based because of the physicality of the artistic medium and, most importantly, conceptual as well because of the symbolic and representational nature of art forms. An artist makes use of concepts as much as a scientist but an artist is also grounded in her physical medium whether that is paint, marble, film or musical sounds. The artist engages both our emotions through our perception of their medium and our intellect through the symbolic representations of their compositions.

The influence of verbal language is not limited to the symbolic representations of the artwork but they also impact on the nature of the artist's perceptive powers, which differs from the perceptive powers of our pre-lingual hominid ancestors. The conceptual powers of the artists change the nature of their perceptive capabilities creating what I have termed secondary perception.

If this hypothesis is correct then there should be a correlation between verbal language and artistic expression as well as evidence for their simultaneous emergence in the history of humankind. Now I must admit in all honesty that this is not a true scientific prediction as I am well aware of the putative correlation of speech and artistic expression claimed by a number of scholars and reviewed below. In addition to my knowledge of this claim I must also credit another source of inspiration for my hypothesis of the influence of secondary perception on artistic expression. That influence is Walter Ong's (1991) notion of secondary orality from which I derived the notion of secondary perception, i.e. perception influenced by verbal language and concept-based thought.

Secondary Perception

In the Extended Mind model (Logan 2007) it was suggested that before language emerged hominid thought was purely percept-based and the brain was basically a percept processor. With the emergence of verbal language the brain bifurcated into the brain and the mind. The brain continued as the seat of percept-based thought and the mind became the seat of concept-based thought. The metaphoric formulation of this notion was captured with the equation: mind = brain + language.

The new insight regarding the emergence of art is that with language and concept-based thought secondary perception emerged in which perception is transformed by conceptual thought in much the same way that orality was transformed by literacy giving rise to secondary orality. Secondary orality is Walter Ong's (1991) simple idea that there is a difference between primary and secondary orality, where primary orality is the orality of a pre-literate culture and secondary orality is the orality of a literate culture. Ong observed that literacy changes the nature of orality creating what he called "secondary orality". Once humans acquired verbal language and conceptual symbolic thought the nature of their perceptual sensorium changed into what we are now calling secondary perception. Secondary perception is to primary or pre-verbal perception what secondary orality is to primary or pre-literate orality. Secondary perception allows the potential artists to combine their perceptual capabilities with their ability to create symbols and to think symbolically, which are the necessary ingredients for artistic expression.

The mechanism by which secondary perception emerges is through downward causation from the mind, the seat of conceptual thought to the brain, the percept processor. This formulation yields a theory for the emergence of art as the product of secondary perception and concept-based thought. It also explains the apparent correlation of the emergence of speech and symbolic art.

Unfortunately this hypothesis can be construed as a just-so story as the emergence of symbolic art and verbal language were each one-time events in the history of humankind. However, one independent prediction is possible based on the idea that artistic expression entails secondary perception and concept-based thought. I believe that a brain scan of an artist composing or performing a work of art would reveal activity in both the part of the brain associated with verbal language and the part of the brain associated with visual or auditory perception depending on the art form. I hope that some experimental psychologists will test this conjecture of mine.

A Google search of the literature revealed that there is a definite impact of the left-brain associated with verbal language skills and conceptual thought on artistic expression, which is largely associated with right brain function.

The following excerpt from a study of an artist who suffered a minor stroke reveals the involvement of both hemispheres in artistic expression indicating that conceptualization plays a role in artistic expression (Annoni et al. 2005, 797).

> Painting is a very complex behavior and its neural correlates involve brain areas processing the perceptive, cognitive, and emotional valences of stimuli; brain damage, therefore, could modify artistic expression.... Right parieto-occipital damage resulting in spatial neglect, constructional apraxia, or perceptual agnosia can alter the spatial configuration of the whole painting or individual parts, while extensive left hemisphere damage may be responsible for simplification of detail of represented objects.

Another neurologist, Anjan Chatterjee (2004, 1573), based on studies of artists with neurological deficits also links conceptualization with artistic expression:

> Thus, from the limited data available, the art of patients with visual agnosias seems to be largely determined by whether their deficit is closer to the perceptual or the conceptual end of object recognition processes. If the deficit is at the perceptual end, patients are likely to not produce the overall form and composition of images, but continue to render individual features of objects. By contrast, patients with deficits at the **conceptual** end are still able to draw very well if copying from a rich source, but fall apart when having to draw from memory or if guided by their knowledge of the world.

Ellen Dissanayake (1988, 112) argues, "the elements of art are human nature's fundamental elements" of which she includes "Language and speech – Classification and concept formation – Symbolization." She goes on to suggest the connection between these elements, "Inseparable from abstract or conceptual thought and language is the ability to symbolize, to recognize one thing as standing for or representing another (ibid., 118)."

The Joint Emergence of Verbal Language and Artistic Expression

A number of scholars have suggested that the emergence of language, symbolic or conceptual thought and artistic expression were all connected and simultaneously began about 50,000 years ago in what Jared Diamond called the "great leap forward" and what Pfeiffer (1982) and Tattersall (1998) call the "creative explosion". It was at this time there was an explosion of human inventiveness when for the first time there emerged a profusion of new tools, clothing made from animal hides, decoration of tools, jewelry, rituals such as ceremonial burials, artistic expression in the form of cave paintings and carved figurines and musical instruments.

> To many archaeologists, art—or symbolic representation, as they prefer to call it—burst on the scene 50,000 years ago, a time when modern humans are widely thought to have migrated out of Africa to the far corners of the globe. These scholars say the migrants brought with them an ability to manipulate symbols and make images that earlier humans had lacked.... As Richard Klein of Stanford University puts it, "There was a kind of behavioral revolution [in Africa] 50,000 years ago. Nobody made art before 50,000 years ago; everybody did afterward." (Appenzeller 1998)

Dunbar (1998, 105) reaches a similar conclusion

> Symbolic language (the language of metaphysics and religion, of science and instruction) would have emerged later as a form of software development (it embodies no new structural or cognitive features not already present in social language), probably at the time of the Upper Paleolithic Revolution some 50,000 years ago when we see the first unequivocal archaeological evidence for symbolism (including a dramatic improvement in the quality and form of tools, the possible use of ochre for decorative purposes, followed in short order by evidence of deliberate burials, art and non-functional jewelry).

Emanuel Anati (1989, 209), an expert on rock art, maintains that art, language and religion have a single root. He also dates the advent of visual art to 50,000 BCE.

> There is no evidence of a full-scale use of visual art until 50,000 years ago. The consistency throughout the world of the same basic repertory of symbols and images exhibited in the early phases of rock art testifies to the common origin of *Homo sapiens* and of his uniquely human intellect.... Early prehistoric men already operated within a framework of mental mechanisms of association, symbolism, and abstraction, which still today are defining characteristics of our species. In comparison to the preceding hominids, using these cognitive skills was not only an evolution, but also a true revolution: a leap forward that once taken has made us forever a very different Primate (Anati 2004)."

David Lewis-Williams (2004), an art historian in his book *The Mind in the Cave: Consciousness and the Origins of Art*, as the title of his book indicates, links the origin of art to consciousness. Like Anati he also links the origin of art to religion. In the Extended Mind model (Logan 2007) that we base this study upon both religion and consciousness depends on conceptual thinking and hence language. We therefore consider the work of Lewis-Williams and Anati as supporting the link between the origin of language and the origin of art.

Similarities and Differences of Verbal Language and Artistic Expression

Both verbal language and artistic expression communicate the intentions of the speaker or writer and the artist and hence both entail a theory of mind, which we will explicate in the next section. Both forms of communication can be used to express emotions. Both require thought and planning although speech and some forms of musical performance tend to be more spontaneous and less planned. But even conversational speech entails a certain amount of planning even though it takes place as the speaker speaks and hence is not a very lengthy process. Both language and art are abstract in that they represent transformations of reality into words or utterances in the case of language and visual forms or sounds in the case of art. As a result both language and art are representational and symbolic. From a Piercian semiotic perspective, however, verbal language is always symbolic but art can be iconic, indexical or symbolic. Iconic representation is representation by similarity as a photograph represents a person. An indexical representation is a sign that is associated with the thing being represented as smoke indexically represents fire. Symbolic representation is when the sign stands for something else by convention. The word "dog" for example represents the four-legged animal that we think of as man's best friend.

Another similarity is the fact that every human culture that we know of possesses both verbal language and artistic expression. They are both universals of the human condition and unique to our species.

> Hominids had been evolving for 4 million years, but art only appeared with *Homo sapiens* and proved to be an exquisitely human expression. The «creation» of art was a revolution
>
> Our ancestor early *sapiens* was characterised by the neurological capacity of creating an ideology, whose basic matrix is still present at the core of modern man's conceptual cognition. This framework included a capacity for synthesis and abstraction, which, among other things, led man to produce art and abstract thought, and to develop an articulate and complex language (Anati 2004, 64 & 60).

And finally to conclude this catalog of similarities it is important to remember that painting, sculpture and music are often referred to as languages and verbal language in the form of oratory, poetry and literature is often referred to as an art form. There are many crossovers between verbal and artistic expression but let us now examine some of the differences.

The arts appeal immediately to the sensual aspect of human thought and then to the intellectual side. Verbal language, on the other hand, appeals immediately to our intellect and then possibly through imagery to the sensual side of our mentality. Verbal language is linear whereas the arts are multidimensional. Even music, which has a temporal linear progression, is multi-dimensional because it is composed of pitch, timbre, tempo, volume, melody and harmony. Verbal language can be analytic and has led to mathematics, science and computing whereas the arts are synthetic and aesthetic. Both verbal language and artistic expression can express both ideas and feelings but the arts tend to be more about feelings and verbal language more about expressing ideas.

A Theory of Mind—Art as Creating an Effect

As was already noted both the arts and verbal language are about communicating intentions and expressing thoughts and feelings. Both forms of communication therefore are based on a theory of mind, i.e. the notion that the communicator believes those who are their audience have a mind similar to their own and hence will comprehend their communication whether that is verbal or artistic. Those who study the origin of language consider the human capability of a theory of mind was a cognitive capability unique to humans that made verbal language possible. Dunbar (1998, 102) defines a theory of mind as "the ability to understand another's individual mental state" without which he claims,

> There would be no language in the form we know it.... Language requires more than the mere coding and deciphering of well-formed

> grammatical statements. Indeed, as has been often pointed out, many everyday conversations are conspicuous by their lack of grammatical structure (Gumperz 1982). However, important formal grammar may be in the precision of information transfer, it is surely the intentionality of speech that is the most demanding feature for both speaker and listener (ibid., 101).

I believe that a theory of mind is just as critical for the origin of artistic expression as it was for the origin of language. Artistic expression is also a uniquely human attribute and also requires a theory of mind mind-set on the part of the artist to be executed. Artists through their artwork are trying to create an effect on their audience and this requires a theory of mind on the part of the artists to believe that they can create effects on their audience like the ones they experience. McLuhan (1964) described the artist's methodology as working backwards from the effect they want to create to the causal elements that will produce the effect they have in mind. In order to work in this manner the artists obviously must have a theory of mind.

Social Communication

Both verbal language and artistic expression are forms of social communication. "Speech...serves two functions, that of social communication, and the representation of and a medium for abstract thought (Logan 2007)." The same may be said of artistic expression. Both verbal language and artistic expression are forms of abstract thought. While both are vehicles for the expression of emotions the visual and musical arts tend to favors emotional expression over analytic thought more so than verbal language. This generalization is only a general trend as one can find superb examples of emotional expression through verbal language and music and visual art that is extremely analytic and everything in between.

Verbal language has been a very important tool for creating social cohesion and cooperation. There is a very strong correlation between altruism and the origin of verbal language. Speech entails the sharing of information which in itself is an altruistic act. Without the desire to help conspecifics there would have been no motivation to want to communicate with fellow humans so there is no doubt that verbal language and altruism go hand in hand. But a similar argument can be made for artistic expression and altruism.

> "Why did humans have the need to record their own thoughts and emotive stimulation? No doubt this is part of the nature of *Homo sapiens*, like socialization, the sense of aesthetic, love, ambition, and solidarity (Anati 2004, 67)."

Art Arising from Mimetic Communication

Merlin Donald has suggested that mimetic communication was the cognitive laboratory in which verbal language developed. The roots of the fine arts can also be traced to percept-based mimetic communication whose basic elements were prosody (the tones of vocalization), facial gesture, hand signals and mime (or body language). The very first art forms were all non-verbal and grew out of mimetic communication. They included music, painting, sculpture and dance all of which were a part of ritual. Music can be traced to the variation of tone and rhythm and hence to the prosody of speech. Dance is basically a form of body language set to music. The first forms of painting were body and face painting and the first forms of sculpture were masks and costumes, which can be seen as attempts to enhance and intensify facial gesture and mime. With the advent of spoken language new hybrid forms of the arts emerged which combined mimetic communication with words to produce modern (post-verbal) art forms such as poetry, which include both words and prosody, songs which combine words and music and theater which combines words with mime and dance (Logan 2007).

Secondary perception not only plays a role in artistic perception but it also contributes technological innovation and design. An inventor of a tool must be able to envisage how the tool will be manufactured and used and hence the inventor's power of perception must combine with his analytic skills.

References

Anati, Emmanuel. 1989. *Les origines de l'art et la fonnation de l'esprit humain.* (Transl.) Diane Ménard (transl.). Paris: Albin Michel.

Anati, Emmanuel. 2004. Introducing the World Archives of Rock Art (WARA): 50.000 years of visual arts. In (XXI Valcamonica Symposium) *New Discoveries, New interpretations, New Research Methods*. Capo di Ponte: Edizioni del Centro.

Anderson, Chris. 2004. The long tail. Wired: October 2004.

Annoni J. M. , G. Devuyst, A. Carota, L. Bruggimann and J. Bogousslavsky. 2005. Changes in artistic style after minor posterior stroke. *Journal of Neurology Neurosurgery and Psychiatry* 76:797–803

Appenzeller, Tim. 1998. Art: Evolution or revolution. Science 282, 1451–54.

Bak, Per. 1996. *How Nature Works*. New York. Copernicus.

Basalla, George. 1988. *The Evolution of Technology.* Cambridge UK: Cambridge University Press.

Bateson, Gregory. 1973. *Steps to an Ecology of Mind.* St. Albans: Paladin Frogmore.

Bertschinger, Edmund. 2000. *The call of science: theological reflections on the ethics of vocation*. Humanity and Cosmos Symposium.

Bickerton, Derek. 1998. Catastrophic evolution: The case for a single step from protolanguage to full human language. In James Hurford, Michael Studdert-Kennedy, Chris Knight (eds.), *Approaches to the Evolution of Language.* Cambridge: Cambridge University Press, 341–58.

Boulding, Kenneth E., 1966, The economics of the coming spaceship earth. In H. Jarrett (ed.), *Environmental Quality in a Growing Economy*, Essays from the Sixth RFF Forum, 3–14. Baltimore, MD: Resources for the Future/Johns Hopkins University Press.

Boyd, Robert. and Peter J. Richerson. 1985. *Culture and the Evolutionary Process.* Chicago: University of Chicago Press.

Brillouin, Léon. 2004. *Science and Information Theory.* Mineola, NY: Dover (First published by Academic Press, 1962).

Brown, Donald E. 1991. *Human Universals.* New York: MacGraw-Hill.

Campbell, Jeremy. 1982. *Grammatical Man: Information, Entropy, Language, and Life.* New York: Simon and Schuster.

Casti, John. 1994. *COMPLEXification.* New York. HarperPerennial.

Chatterjee, Anjan. 2004. The neuropsychology of visual artistic production. *Neuropsychologia* 42 (2004) 1568–1583.

Christiansen, Morten. 1994. Infinite languages finite minds: Connectionism, learning and linguistic structure. Unpublished doctoral dissertation, Centre for Cognitive Studies, University of Edinburgh UK.

Christiansen, Morten. 1995. Language as an organism – implications for the evolution and acquisition of language. Unpublished manuscript, Washington University.

Christiansen, M; Dale, R.; Ellefson, M.; and Conway C. 2001. The role of sequential learning in language evolution: computational and experimental studies. In: Cangelosi, A. and Parisi, D. (eds.), *Simulating the Evolution of Language.* London: Springer-Verlag.

Christiansen, M.& M. Ellefson. 2002. Linguistic adaptation without linguistic constraints: The role of sequential learning in language evolution. In A. Wray (ed), *The Transition to Language.* Oxford: Oxford University Press, 335–58.

Chomsky, Noam. 1957. *Syntactic Structures.* The Hague: Mouton.

Clark, Steven. 2000. Private communication.

Clayton, Philip. 2004. *Mind and Emergence: From Quantum to Consciousness.* Oxford: Oxford University Press.

Cohen, Jack and Ian Stewart. 1994. *The Collapse of Chaos: Discovering Simplicity in a Complex World.* New York. Penguin Books.

Coyne, George. 2000. When the sacred cows of science and religion meet. Humanity and Cosmos Symposium.

Crow, T. J. 2002. Candidate gene for cerebral asymmetry. In A. Wray (ed.), *The Transition to Language.* Oxford: Oxford University Press, 93–112.

Cziko, Gary. 1995. *Without Miracles: Universal Selection Theory and the Second Darwinian Revolution.* Cambridge MA: MIT Press.

Dawkins, Richard. 1989 edition (originally published in 1976). *The Selfish Gene.* Oxford: Oxford University Press.

Dawkins, Richard. 1996. The survival machine. In Brockman, John (ed), *The Third Culture.* New York: Touchstone Books.

Deacon, Terrence W. 1997. *The Symbolic Species: The Co-evolution of the Brain and Language.* New York: W. W. Norton & Co.

Dissanayake, Ellen. 1988. *What is Art For?* Seattle: University of Washington Press.

Donald, Merlin. 1991. *The Origin of the Modern Mind.* Cambridge, MA.: Harvard University Press.

Dunbar, Robin. 1998. Theory of mind and the evolution of language. In James Hurford, Michael Studdert-Kennedy, Chris Knight (eds), *Approaches to the Evolution of Language.* Cambridge: Cambridge University Press, 92–110.

Durham, William H., 1991. *Coevolution: Genes, Culture and Human Diversity.* Stanford: Stanford University Press.

Eldredge, Niles and Stephen Jay Gould, 1972."Punctuated equilibria: an alternative to phyletic gradualism. In T.J.M. Schopf, ed., *Models in Paleobiology.* San Francisco: Freeman Cooper. 82–115.

El-Hani, Charbel Nino and Antonio Marcos Pereira. 2000. Higher-level descriptions: why we should preserve them? In Peter Bogh Andersen, Claus Emmeche, Niels Ole Finnemann and Peder Voetmann Christiansen (eds). *Downward Causation; Mind, Bodies, and Matter.* Aarhus: Aarhus University Press, 118–42.

Fisher, R.A. 1925. Theory of statistical estimation. *Proceedings of Cambridge Philosophical Society* XXII, 709.

Frohmann, Bernd. 2004. *Deflating Information.* Toronto: University of Toronto Press.

Geertz, Clifford. 1973. *The Interpretation of Culture.* New York: Basic.

Goodenough, Ward. 1981. Culture, *Language and Society. McCaleb Module in Anthropology.* Menlo Park: Benjamin/Cummings.

Gould, Stephen J. 1996. The pattern of life's history. In Brockman, J. (ed.), *The Third Culture.* New York: Touchstone Books.

Gozzi, Raymond. 2000. Private communication by email.

Gumperz, J. J., 1982. *Discourse Strategies*. Cambridge: Cambridge University Press.

Hartley, R.V.L. 1928. Transmission of information. Bell Systems Technical Journal VII: 535–63.

Hayles, Katherine. 1999a. *How We Became Posthuman*. Chicago: University of Chicago Press.

Hayles, Katherine. 1999b. The condition of virtuality. In Peter Lunenfeld (ed.), *The Digital Dialectic*. Cambridge MA: MIT Press.

Hood, L. and D. Galas. 2003. The digital code of DNA. *Nature* 421: 444–448.

Hudson, R. 1984. *Word Grammar*. London: Basil Blackwell.

Johnson, Allen W. and Timothy Earle. 1987. *The Evolution of Human Societies: From Foraging Group to Agrarian State*. Stanford: Stanford University Press.

Kauffman, Stuart. 1995. *At Home in the Universe*. Oxford: Oxford University Press.

Kauffman, Stuart. 2000. *Investigations*. Oxford: Oxford University Press.

Kauffman, Stuart. 2008. *Reinventing the Sacred*. New York: Basic Books.

Kauffman, Stuart A. and Philip Clayton. 2006. On emergence, agency, and organization. Biology and Philosophy 21:501–521.

Kauffman, Stuart, Robert K. Logan, Robert Este, Randy Goebel, David Hobill and Ilya Shmulevich. 2007. Propagating organization: an enquiry. *Biology and Philosophy* 23: 27–45.

Kelly, Kevin. 1994. *Out of Control: the New Biology of Machines, Social Systems and the Economic World*. Boston: Addison-Wesley, 98.

Kortlandt, Frederik. 2003. The origin and nature of the linguistic parasite. In Brigitte Bauer and Georges-Jean Pinault (eds.) *Language in Time and Space: A Festschrift for Werner Winter on the Occasion of his 80th Birthday*. Berlin: Mouton De Gruyter,

Kroker Arthur. 2000. Code of Privilege. An online interview by Sharon Grace.

Kuhn, Thomas. 1972. *The Structure of Scientific Revolutions*. Chicago: University of Chicago Press.

Langefors, B. 1968. *System för Företagsstyrning*. Lund: Studentlitteratur.

Lee, N. and Schumann, J.H. 2003. The evolution of language and the symbolosphere as complex adaptive systems. Paper presented at the conference of the American Association for Applied Linguistics, Arlington VA, March 22–25.

Lewis, Gilbert N. 1930. The symmetry of time in physics. *Science* 71: 569–576

Lewis-Williams, David. 2004. *The Mind in the Cave: Consciousness and the Origins of Art*. London: Thames and Hudson.

Logan, Robert K. 1979. The Mystery of the Discovery of Zero. *Etcetera* 36: 16–28.

Logan, Robert K. 1995. *The Fifth Language: Learning a Living in the Computer Age*. Toronto: Stoddart Publishing.

Logan, Robert K. 1997. The extended mind: understanding language and thought in terms of complexity and chaos theory. Presented at the 7th Annual Conference of The Society for Chaos Theory in Psychology and the Life Sciences at Marquette U., Milwaukee, Wisconsin, Aug. 1, 1997.

Logan, Robert K. 2000. The extended mind: understanding language and thought in terms of complexity and chaos theory. In Lance Strate (ed.), 2000 *Communication and Speech Annual* 14. New York: The New York State Communication Association.

Logan, Robert K. 2002. The five ages of communication. *Explorations In Media Ecology* 1:13–20.

Logan, Robert K. 2004a. *The Alphabet Effect*. Cresskill NJ: Hampton. 1st edition 1986. New York: Wm. Morrow.

Logan, Robert K. 2004b. *The Sixth Language: Learning a Living in the Internet Age*. Caldwell NJ: Blackburn Press. 1st edition 2000. Toronto: Stoddart Publishing.

Logan, Robert K. 2004c. *Collaborate to Compete*. Toronto: Wiley Canada

Logan, Robert K. 2005. Neo-Dualism and the bifurcation of the symbolosphere into the mediasphere and the human mind. *Semiotica* 157 (1/4), 345–51.

Logan, Robert K. 2006a. The extended mind model of the origin of language and culture. In Nathalie Gontier, Jean Paul Van Bendegem and Diederik Aerts (eds), *Evolutionary epistemology, language and culture*. Dordrecht: Springer.

Logan, Robert K. 2006b. Neo-Dualism and the bifurcation of the symbolosphere into the mediasphere and the human mind. *Semiotica* 160 (1/4), 229–42.

Logan, Robert K. 2007. *The Extended Mind: The Origin of Language and Culture*. Toronto: University of Toronto Press.

Logan, Robert K. and Schumann, John. 2005. The symbolosphere, conceptualization, language and neo-dualism. *Semiotica* 155: 201–14.

Logan, Robert K. and Louis W. Stokes. 2004. *Collaborate to Compete: Driving Profitability in the Knowledge Economy.* Toronto and New York: Wiley

Losee, Robert M. 1997. A discipline independent definition of information. *Journal of the American Society for Information Science* 48 (3): 254–269.

MacArthur, Daniel. 2000. Cosmology, creation, and psuedo questions. Humanity and Cosmos Symposium.

MacKay, Donald M.. 1969. *Information, Mechanism and Meaning*. Cambridge MA: MIT Press.

Margulis, Lynn. 1970. *Origin of Eukaryotic Cells.* Yale University Press.

Maturana, Humberto and R. Varela. 1992. *The Tree of Knowledge.* Boston: Shambala.

McLuhan, Marshall. 1954. New media as political forms. *Explorations* 3 (Aug.): 120–126.

McLuhan, Marshall. 1962. *The Gutenberg Galaxy.* Toronto: Univ. of Toronto Press.

McLuhan, Marshall. 1964. *Understanding Media*. New York: McGraw Hill. *The page references in the text are for the McGraw Hill paperback second edition. Readers should be aware that the pagination in other editions is different. To aid the reader in calibrating, note that Chapter 1, The Medium is the Message, begins on page 7 in the edition I have referenced —RKL.*

McLuhan, Marshall. 1975. Communication: McLuhan's laws of media. *Technology and Culture* 16 (1): 74–78

McLuhan, Marshall. 1995. A McLuhan Sourcebook assembled by William Kuhns. In Eric McLuhan and Frank Zingrone (eds.) *Essential McLuhan.* Concord, ON: Anansi, 272 & 276.

McLuhan, Marshall. 2004. *Understanding Me: Lectures and Interviews.* Stephanie McLuhan and David Staines (eds). Cambridge MA: MIT Press.

McLuhan, Marshall, and McLuhan, Eric. 1988. *Laws of Media: The New Science.* Toronto: University of Toronto Press.

McLuhan, M. and R. K. Logan. 1977. Alphabet, mother of invention. *Et Cetera,* 34, 373–383.

Mokyr, Joel. 1990. *The Lever of Riches: Technological Creativity and Economic Progress.* New York: Oxford University Press.

Olesen, Mogens. 2008. Sketching an evolutionary understanding of digital media. Department of Film and Media Studies, U. of Copenhagen. Unpublished paper.

Ong, Walter. 1991. *Orality and Literacy: The Technologizing of the Word.* London: Routledge Kegan & Paul

Pfeiffer, John. 1982. *Creative Explosion.* New York: Harper & Row.

Popper, Karl. 1959. *The Logic of Scientific Discovery.* (originally published in German as *Logik der Forschung*). London: Routledge.

Popper, Karl. 1979. *Objective Knowledge: An Evolutionary Approach* (rev. ed.). Oxford: Clarendon Press, 261.

Prigogine, Ilya and I. Stengers. 1984. *Order Out of Chaos.* New York: Bantam Books.

Prigogine, Ilya. 1997. *The End of Certainty.* New York: Free Press.

Sarkar, Sahotra. 1996. Decoding "coding" — information and DNA. *Bioscience* 46 (11): 857–64.

Schrödinger, Erwin. 1992. *What is Life?* Cambridge: Cambridge University Press.

Schmandt-Besserat, Denise. 1986. The origins of writing. *Written Communication* 13.

Schneider, Eric and Dorion Sagan. 2005. *Into the Cool: Energy, Flow, Thermodynamics and Life.* Chicago: University of Chicago Press.

Schumann, John H. 2003a. The evolution of language: What evolved? Paper presented at the Colloquium on Derek Bickerton's Contributions to Creolistics and Related Fields, The Society for Pidgin and Creole Linguistics Summer Conference, Aug. 14–17, University of Hawaii, Honolulu.

Schumann, John H. 2003b. The evolution of the symbolosphere. Great Ideas in the Social Sciences Lecture, UCLA Center for Governance, Nov. 21.

Searchnetworking.com. 2002. Information theory. http://searchnetworking.techtarget.com/sDefinition/0,,sid7_gci801374,00.html

Shannon, Claude E. 1948. A mathematical theory of communication. *Bell System Technical Journal* 27, 379–423 and 623–656, July and October, 1948.

Shannon, Claude E. 1953. The lattice theory of information. *Information Theory, IEEE Transactions* 1:105–07.

Shiryayev, A.N. (Editor). 1993. *Selected Works of A.N. Kolmogorov: Volume III: Information Theory and the Theory of Algorithms (Mathematics and its Applications).* New York: Kluwer Academic Publishing.

Tattersall, Ian. 1997. *From Becoming Human: Evolution and Human Uniqueness.* New York: Harcourt.

Taylor, Mark. 2002. *The Moment of Complexity: Emerging Network Culture.* Chicago: University of Chicago Press.

Tomasello, Michael. 1999. *The Cultural Origins of Human Cognition.* Cambridge, MA: Harvard University Press.

Tomasello, Michael, A. C. Kruger, and H. H. Ratner. 1993. Cultural learning. *Behavioral and Brain Sciences* 16: 495–552.

Tzannes, Nicholas S. 1968. The concept of 'meaning' in information theory. (August 7, 1968) In Warren McCulloch Papers, American Philosophical Society Library, Philadelphia, B/M139, Box 1 (cited in Hayles 1999a, 56 & 301 n. 7).

Van Alstyne, Greg and Robert K. Logan. 2007. Designing for emergence and innovation: Redesigning design. *Artifact* 1: 120–29.

van Driem, George. 2005. The language organism: the Leiden theory of language evolution. In James W. Minett and William S.-Y. Wang (eds) *Language Acquisition, Change and Emergence: Essays in Evolutionary Linguistics.* Hong Kong: City University of Hong Kong Press.

Vincenti, Walter. 1990. *What Engineers Know and How They Know It.* Baltimore: Johns Hopkins University Press.

von Bertalanffy, Ludwig. 1968. *General Systems Theory: Foundations, Development, Applications.* New York: George Braziller.

Vygotsky, Lev. 1962. *Thought and Language.* Cambridge MA: MIT Press.

Wicken, Jeffery. 1987. Entropy and information: Suggestion for a common language. *Philosophy of Science* 54: 176–193.

Wiener, Norbert. 1948 (2nd ed., 1961). *Cybernetics; or, Control and Communication in the Animal and the Machine.* Cambridge MA: MIT Press.

Wiener, Norbert. 1950. *The Human Use of Human Beings.* Boston: MIT Press.

Keywords

This list of keywords is offered in lieue of an index for locating concepts, ideas and proper noun references. Readers of the digital edition of this text, available via demopublishing.com, may select a term below or enter any other term to search for occurances.

Colophon & Image Credits

What is Information? was designed by Greg Van Alstyne and Garry Ing.
Print on demand (POD) and distribution organized with the generous guidance of Nimble Books.
Typefaces: Skolar (2007), designed by David Březina, published by Rosetta.
Etica (2000), designed by Leftloft (Milan), published by TypeTogether.

Cover collage by Greg Van Alstyne with image elements courtesy Kwamikagami; Jer Thorp; Vossman; Edmund B. Wilson; Zephyris, under Creative Commons license. Chapter introductory images courtesy their respective authors via Flickr.com, under Creative Commons license.

Image, Source:
Stuart Kauffman, Teemu Rajala
Schroedinger, Public Domain, 1933
Claude Shannon, it-science.net
Norbert Wiener, Konrad Jacobs
Macy Conference participants, www.asc-cybernetics.org
Gregory Bateson, Eliot Elisofon
R. A. Fisher, Public Domain
Maxwell, G. J. Stodart
Leo_Szilard, U.S. Department of Energy, Pubic Domain
Cell, National Center for Biotechnology Information
Cells in culture, John Schmidt
DNA, RNA and a protein, Zephyris
RNA, Vossman
Protein, Splette
Katherine Hayles, Dave Pape
Marshall McLuhan, John Reeves
Marshall McLuhan and Robert K. Logan, Maria Ielenszky Logan
Clay token, Marie-Lan Nguyen
Images of clay envelopes, Marie-Lan Nguyen
Clay tablet, Jastrow
Mesopotamian letters, Semitic, Greek, and Roman alphabets, Kwamikagami
Proto-Sinaitic script, Zander Schubert
Alphabets and Semitic scripts, Wikipedia
Hindu, Arabic and Roman numerals, abacus, Madden, GFDL
Parminides, Wikipedia
Painting of Adam and Eve, Public Domain
Hunter gatherer, Forrester's Pictorial Miscellany for the Family Circle, 1855, Public Domain
Egyptian farming, The Yorck Project: 10.000 Meisterwerke der Malerei
Biosphere, NASA/Apollo 17 crew taken by either Harrison Schmitt or Ron Evans
Noam Chomsky, Duncan Rawlinson
Religious symbols, Wikipedia
Alphabets, Wikipedia
Crab Nebula, NASA, ESA, J. Hester and A. Loll, Arizona State University
Photograph of telescope, Andrew Dunn
Physics collage, Wikipedia
Isaac Newton, Portsmouth Estate, 1689
Portrait of Isaac Newton, William Blake
Gutenberg Bible, Kevin Eng
Printing press, Philip B. Meggs

Acknowledgments

I wish to thank my colleagues in sLab at OCAD University, the editorial and design team at DEMO Publishing for producing such a beautiful book, and the scholars who helped me develop the ideas in this book including Stuart Kauffman, John Schumann, Greg Van Alstyne and Terrence Deacon. Finally I wish to thank my family for their patience including my wife Maria and my children Renee, David, Natalie, Rebecca, and Paul.
—Robert K. Logan

We would like to extend our sincere appreciation to Lenore Richards, Program Director of OCAD University's Master of Design in Strategic Foresight & Innovation, for providing research seed funding to help in launching this, our first publication. We would also like to thank Fred Zimmerman, publisher at Nimble Books, for his generous assistance to DEMO Publishing.
—Peter Jones and Greg Van Alstyne, the editors

About the Author

Robert K. Logan is Chief Scientist of Strategic Innovation Lab (sLab) at OCAD University; Professor Emeritus of Physics at the University of Toronto; and Fellow of the University of St. Michael's College (U of T). He has published over one hundred articles and twelve books in fields including physics, media ecology, linguistics, management, politics, and information theory. Details of his work can be found online at: https://utoronto.academia.edu/RobertKLogan and http://www.physics.utoronto.ca/Members/logan.

Socialize this book!

**Design
Emergence
Media
Organization**

DEMO Publishing presents new genres and formats of interdisciplinary design and media research. We seek to illuminate complex social and organizational systems, to foster foresight and innovation. All of these domains manifest emergence and demand continual updating of our systemic design thinking.

One of our goals is to explore the social dimension of reading, writing and publishing. To further our research in this regard we are simultaneously publishing this book as both a printed book and an ebook under a **Creative Commons Attribution-NonCommercial (CC BY-NC) license**. This license lets others build upon or remix the work non-commercially. Although new works must acknowledge the original source and be non-commercial, derivative works can be licensed on new terms. In all derivative works of text, the author requires attribution.

The ebook is available at demopublishing.com. Via our website readers will find features to create comments and foster dialogue with each other and the author. Dr. Logan also invites personal comments directly at his University of Toronto email address: **logan@physics.utoronto.ca**.

We hope that you will join us in exploring the social interaction with a book's ideas and its readership that the Internet makes possible. A book is no longer a closed package of information but rather a social and intellectual process that includes all the book's reviews, all of the comments and all of the references made to the book. We look forward to interacting with you online!